AMERICAN WATCHMAKING

A Technical History

of the

American Watch Industry

1850-1930

by

Michael C. Harrold

i

Library of Congress Cataloging-in-Publication Data
Harrold, Michael C., author.
American watchmaking : a technical history of the American watch industry,
1850-1930 / by Michael C. Harrold.
Columbia, PA : National Association of Watch and Clock Collectors, Inc.,
2017. | "A Supplement to the Bulletin of the National Association of Watch
and Clock Collectors, Inc. P.O. Box 33, Columbia, PA 17512 Number 14,
Spring 1984." | Reprint. | Includes index.
LCCN 2016054343| ISBN 9781944018030 (pbk. : alk. paper) | ISBN
1944018034 (pbk. : alk. paper)
LCSH: Clock and watch making--United States--History.
LCC TS543.U6 H28 2017 | DDC 681.1/130973--dc23
LC record available at https://lccn.loc.gov/2016054343

2nd printing April 2017
©1981 M.C. Harrold

Printed in the United States of America
The National Association of Watch & Clock Collectors, Inc.
514 Poplar Street, Columbia, PA 17512
2nd Printing Editors: Therese Umerlik and Christiane Odyniec
2nd Printing Creative Services Production Leader: Keith Lehman

Illustrations by Robert Backstrom
Photographs and cuts courtesy of Roy Ehrhardt
Miscellaneous period cuts from
Kenneth D. Roberts Publishing Co. and Adams Brown Co.
Originally published as a Supplement to the BULLETIN of the
National Association of Watch & Clock Collectors, Inc.
P.O. Box 33, Columbia, PA 17512
Number 14, Spring 1984

About the Covers

Taken in 1910 at the Waltham Watch Company, the front cover depicts men and women
working together on an assembly floor. Known as the "Waltham System," the company
had an ethos that prioritized the safety and well-being of its employees and recognized
women as important and equal members of its workforce. The postcards on the back cover
show the following watch companies, from top to bottom, left to right:
E. Howard & Co., Dueber-Hampden Watch Company, Waltham Watch Company,
Hamilton Watch Company, and Elgin National Watch Company.

My father said that anything is interesting if you bother to read about it.

TABLE OF CONTENTS

Figures and Tables

FOREWORD

With this work, Mike Harrold has accomplished what has desperately been needed by watch enthusiasts for many years . . . a comprehensive overview of American watchmaking in a single reference source. Until now, this task had not been effectively tackled, with the result that a myriad of books, booklets, articles, monographs, and other miscellaneous references were required reading for anyone wanting a general grasp of American watchmaking.

For some 100 years, the classic works of Crossman and Abbott have served as primary references. The major problem with these two works is that they were completed relatively early in the life cycle of American watchmaking when today's history was still in the future. Since then there have been a number of excellent efforts to document events in the field, but these have largely been specialized, concentrating on the "trees" rather than the "forest."

Another problem has been the overwhelming complexity of the "forest." Most watch historians have chosen to specialize and have not been inclined, or qualified, to undertake a monograph on the broader aspects of the American watchmaking industry. Mike Harrold has met the challenge head on, covering for an 80-year period from 1850 to 1930 both technical developments of watches and their manufacture, and trends in the business aspects of the industry. In doing so, the life span of the industry has been broken into four stages, from birth to death, and watches have been grouped into three broad classes. Such categorizing entails the risk of channeling people's thinking with artificial constraints. The risk is justified in this case, for out of the forest of details a path has been charted on which one can walk through the life and times of American watchmaking. Such categories have given sensible form to a full range of the industry, its companies and products, and individuals who worked both in the factories and for themselves.

In addition to excellent photographs, Mike has illustrated his work with many significant graphics, including charts, tables, line graphs, and schematics which visually assist his development of key points concerning American watch history. These graphics are a major contribution, as they provide heretofore unpublished trends in the development of American watches. His work is well documented and a wide variety of references are listed for those wanting additional information.

Mike Harrold is to be commended for a job well done . . . he modestly calls his work "a primer" of American watch history. This effort then is one of the most advanced primers it has been my good fortune to read concerning the rise and fall of the American watchmaking industry.

Eugene T. Fuller
FNAWCC, Research Committee
February, 1984

AMERICAN WATCHMAKING

A Technical History of the American Watch Industry 1850-1930

by Michael C. Harrold

Preface

People are often drawn to a watch by some physical attribute such as a beautiful case, number of jewels, or a clever escapement. There are also such indirect attractions as scarcity and dollar value. These all revolve around the particular timepiece, so that the aficionado expends considerable effort learning of individual watches and their features. That is natural since anyone wishing to purchase a watch must know its nature and its worth. Beyond this, there may also be a desire to understand how the nature of that watch evolved. Through the centuries, watchmaking was influenced by many factors, and was itself an influence in shaping other events. Navigational timekeepers from London watchmaking shops contributed to the naval power that supported the British empire. In the United States, watchmaking was among the industries that changed America from a frontier source of natural resources to a modern supplier of technology.

The history of American watchmaking has been difficult for people to assimilate. Most published information concerns particular watches or companies, concentrating on hardware rather than its development. Much historical information that is available is reprinted from contemporary accounts, which assume the reader is aware of many unspoken details and then-current events that have long been forgotten. Piecing together a story from such scattered ephemera consumes time and effort, often yielding meager or misleading results. The reader is left to incorporate scraps of information into a general background. Trends and conclusions must then be reached before this swirling mass of names, places, times, and companies can be properly correlated to an individual watch. The payoff for this effort is knowledge of the forces that shaped the watch industry and its products.

This book is intended as an aid in understanding how American watches developed. It is not an encyclopedia but a primer of watch history. Once primed, the reader will be better able to form his own conclusions from both available timepieces and printed information. He may eventually discover that not only does history illuminate watches, but watches provide peepholes into history.

M.C.H. 1981

Each of us is the sum of our experience. If anyone is helped by this book, it is only because of the help and encouragement I have received from others. In my personal experience, Ralph Warner and Jesse Medlock must be thanked for directing my earliest years of watch collecting. Credit goes to Dick Ziebell and Paul Wing for innumerable hours of discussions, often raising more questions than answers, that have kept my interest alive. Likewise, Gene Fuller has provided many bits in considering American watch history. Also, my gratitude to Roy Ehrhardt for allowing me to tap his vast library of photographs. Most importantly, praise must go to the many before me who have had the energy and fortitude to commit themselves to publication; without them, few of us would know anything.

Introduction To American Watchmaking

The history of American watchmaking is a story of the factory system producing interchangeable parts on automatic machinery. It was among the earliest industries to carry these techniques to a high level, supported by demand for a moderately priced watch; for while the industrial revolution was raising living standards in 1850, the average person could not yet afford a watch. Prices of imports were dropping in the years just before the industry developed, especially Swiss watches, which were consistently less expensive than finer English timepieces. An English rack lever might have cost $50 in the 1820's but by 1850 an inexpensive English lever watch was $30 while the Swiss lever and cylinder watches might have been down to $15. These were still prodigious prices in their day. America was largely agricultural, so many people did business by barter and helpers were often hired for room and board, but no pay. Under such circumstances any price at all was beyond the average grasp. Even in industrial areas where wages were paid, the price of labor was cheap. In 1850 a worker made less than a dollar a day, ten hours a day, six days a week. (This was still expensive labor, for European workers earned even less.) By 1900 the work week was down to 50 hours and daily wages were approaching $1.50. The working middle class had been created and they would buy a watch if somebody would build one in their price range.

1

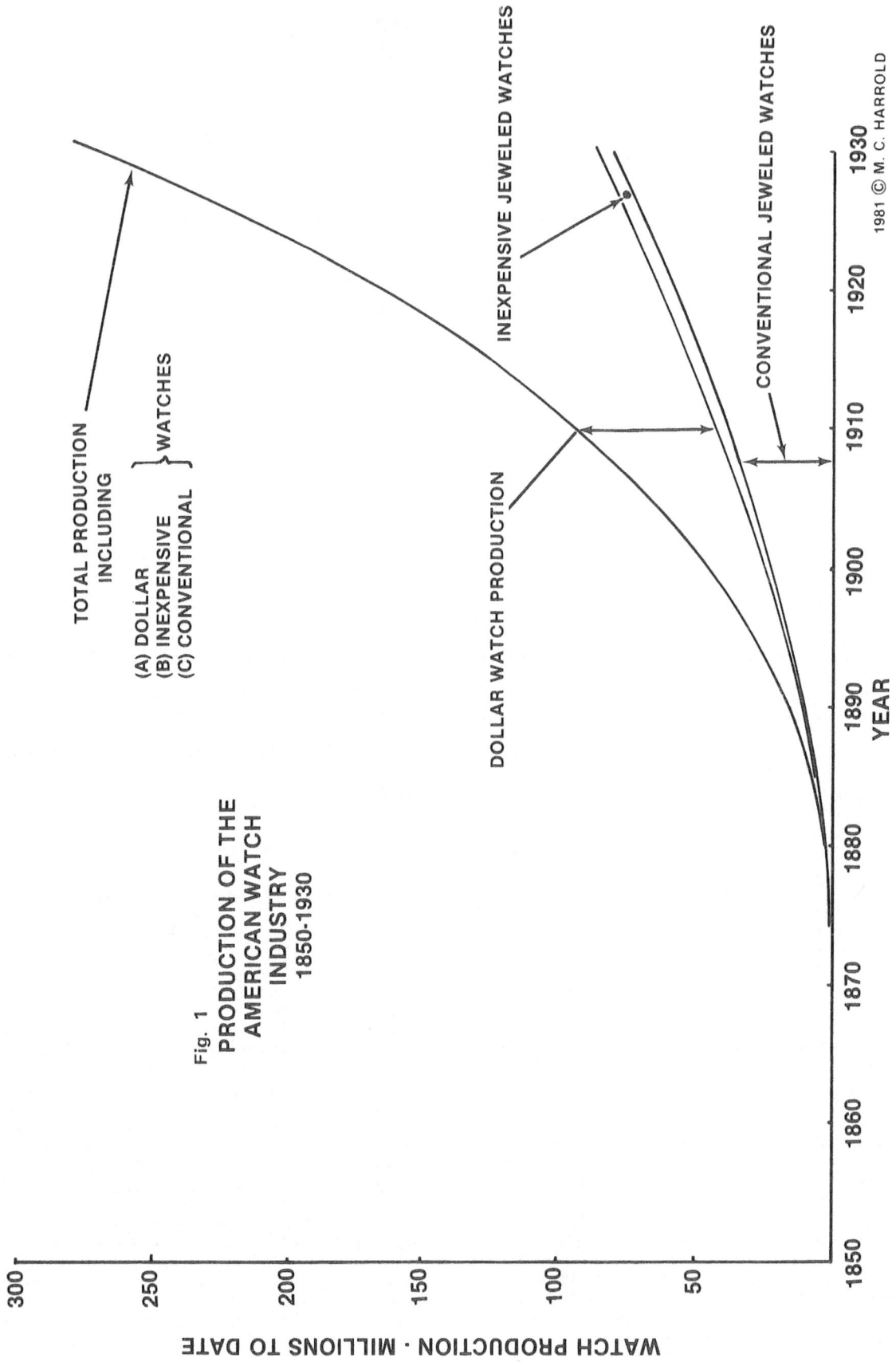

Fig. 1
PRODUCTION OF THE
AMERICAN WATCH
INDUSTRY
1850-1930

TOTAL PRODUCTION
INCLUDING

(A) DOLLAR
(B) INEXPENSIVE } WATCHES
(C) CONVENTIONAL

INEXPENSIVE JEWELED WATCHES

CONVENTIONAL JEWELED WATCHES

DOLLAR WATCH PRODUCTION

1981 © M. C. HARROLD

WATCH PRODUCTION · MILLIONS TO DATE

YEAR

2

In 1850 the European craft system was building watches which few people could afford. There were some within the crafts who saw the possibility of watchmaking by machinery but their efforts were mostly unsuccessful since machinery was resisted by the craft guilds. America had no established guilds to block the rise of the factory system. Visionaries here were free to develop techniques for making large quantities of watches for the ever increasing number of people who could afford them. Factories became the American way, bringing the average citizen farm implements, carriages, tools, guns, sewing machines, clothes, clocks, watches, and an almost endless list of other goods, through machine technology.

When the American watch industry began, in 1850, it had difficulty finding acceptance, and only after fifteen years of effort did fortunes improve. Prosperity then came to watch factories, and over the next two decades promoted growth of a sophisticated manufacturing system. In response to increasing demand, three classes of watches developed within American watchmaking:

1) conventional jeweled watches;
2) inexpensive jeweled watches;
3) dollar watches.

This progression represented a shift in watch status from an exclusive possession to a common item.

Conventional jeweled watches supported development of the industry. Much of their overall appearance was initially derived from English watch designs but a distinctly American character soon evolved. Though they ranged down to seven jewels they were made as quality timepieces. For the most part they were adjusted and temperature compensated, built for long life, and designed to be repairable. Manufacturing cycles were consequently longer than with lesser types of watches and prices were correspondingly higher.

Inexpensive jeweled watches spanned a wide range between conventional quality timepieces and dollar watches. The intent was to make these slightly cheaper than the lowest priced conventional watches in order to attract a larger market. They used lower grade materials and finish, so that performance was also poorer, yet they could provide adequate timekeeping for a considerable period. Since sophisticated equipment was required to produce such watches at low cost, some years passed before they developed. In fact, companies specializing in inexpensive jeweled watches did not appear till 1883, later than the dollar watch. Even then such factories lived a compromised existence for they made neither the best nor the cheapest watches. Therefore only a few companies succeeded in this area as a fulltime business.

The unjeweled dollar watch was the logical limit in reducing the cost of a useful pocket watch. It must be remembered that dollar watches were intended as serious timekeepers, not throwaway items. Building dollar watches clearly depended on making cheap interchangeable parts that could be assembled just as they came off the machines. Though minimum quality materials and manufacturing processes were used, the watches lasted for several years and were cheaper to repair than replace. Dollar watches appeared in 1878 and the business rapidly boomed, indicating the massive demand for a low-cost timepiece.

Figure 1 shows total output for the American watch industry through 1930. Manufacture of conventional jeweled watches began in 1850 but more than 25 years elapsed before significant quantities were produced. By then the dollar watch had begun its rapid acceleration. Dollar watch output soon shot ahead of jeweled watch production and remained there. In contrast, inexpensive jeweled watchmaking saw a number of its companies fail, and the few

survivors were never really significant producers. It was difficult for them to compete head-on against the cheapest models from the major manufacturers of conventional watches, and most people were buying dollar watches anyway. The unmistakable trend was the spread of low-cost watches throughout an increasingly mechanized industry.

The active period of the American watch industry ran from 1850 to 1930. Rounding dates, this lifespan can be resolved into four distinct stages:

1) 1850-1860 Development;
2) 1860-1880 Expansion;
3) 1880-1910 Competition;
4) 1910-1930 Decline.

Figure 2 is a map of watch companies through these years. The four phases become apparent by considering changes in birth and mortality rates.

The developmental period (1850-1860) was characterized by a few pioneers establishing the principles of machine watch manufacture. This was a typically difficult learning stage with some conspicuous failures, but it succeeded in attracting attention. Consequently, Northern prosperity during the Civil War ushered in the period of expansion (1860-1880) with a gush of new companies in 1864. New company formation continued through this period, and more important, manufacturing capability expanded at even a greater rate. After the Civil War, wholesale prices and business conditions suffered a general decline till the turn of the century. However, the watch industry enjoyed an era of growth since it was entering a previously unoccupied position in the marketplace. Watches had long been a status symbol and people were eager to own one as prices dropped within reach. A nationalistic spirit also supported American products. As a result, early companies could set prices with little pressure from competition, and were concerned mostly with making and distributing movements quickly enough.

The United States watch industry had accomplished a great deal in its first 30 years. There was little previous experience with machine manufacture so the industry developed everything as it went along. Though the small size of watch parts presented unique difficulties, it was also of some advantage: watchmaking machines were correspondingly compact. Eventually, complex automated systems could be assembled in a single room and a moderately sized factory could house a virtual industrial army. It was this capability that pushed watchmaking past other fields of manufacture and dropped the price of an American jeweled watch from $40 in 1850 to $10 in 1880.

The Philadelphia Centennial Exposition of 1876 provided a perfect opportunity for presenting these accomplishments to the world. Waltham and other watch companies impressively demonstrated a number of their machines plus their finest watches, and it became clear that American factories could saturate most segments of the market with well-made timekeepers. As if to demonstrate the point, the industry unleashed their unjeweled watch in 1878 for $3.50, tearing away the low-priced market that had been the last Swiss stronghold. By the turn of the century these were down to $1, which European handcrafts could never have accomplished. The same capability was there when railroads demanded large numbers of high quality timekeepers. These rolled out by the thousands at $20 to $40, a price trainmen could afford. On top of all this the industry provided an entirely new service by pouring out a river of spare parts. Whether a person carried a dollar watch or one of the industry's finest creations, quality parts were available everywhere in the country so that it could be repaired quickly and accurately by workmen of average skill. Even the most notable foreign competition could not match this since they could not produce interchangeable finished parts. Whereas the Swiss controlled much of the watch market in 1840, especially in products under $40, they did less American business in 1880 than the Waltham plant alone. Swiss watches were even being tailored to look like popular American movements to boost sales. The takeover by machine-made watches was complete, with dollar-type watches sweeping the low price range while the finest pieces from Howard, Waltham and Elgin brought over $150.

Needless to say, carefree expansion and unbounded opportunity could not last forever and the period of competition (1880-1910) grasped the industry with a force that never completely relaxed. Significant parts of the market had already been tapped and there was a generally declining business trend through the last of the century. Additional effort was therefore required to increase sales, and watch companies bumped into each other at every turn. Many new companies entered watch manufacture based on the previous era of prosperity but found an atmosphere that scarcely nurtured growth. Prices were being cut across the trade and successful factories were dipping into reserves to finance efficient production methods. Even the promising field of dollar watches became rapidly overcrowded. This kept young companies struggling while weaker firms wilted under the pressure. Economic recessions during the 1890's added to existing burdens, triggering a rapid series of business failures that extended into the 1900's. After the turn of the century, the last new watch companies were formed and jeweled watch sales became stagnant.

As the Twentieth Century began moving, watch manufacturing was no longer vibrant and decline set in (1910-1930). Companies found existence burdensome and were willing to sell out in the unlikely event a buyer could be found. Economic fluctuations continued to hack away at the dwindling number of watchmakers, collapsing the large Ingersoll business and causing a complete reorganization of the Waltham plant. While the 1920's proved prosperous to much of the country, problems continued to plague American watchmaking. The Swiss were back. They had been diligently reorganizing their industry, adopting machine manufacture, and learning modern marketing. They had also been placing considerable emphasis on the tiny wristwatch and were successfully increasing its popularity after World War I. Meeting demand for wristwatches meant investing in new production equipment, an expensive proposition. Few companies had sufficient cash reserves, and credit was difficult to find since troubles at Ingersoll and Waltham had frightened the financial community away from watchmaking. To make matters worse, the small size of wristwatches presented new technical difficulties for precision manufacture, just the sort of problem nobody needed. So the 1920's were another period in which numerous firms disappeared from watchmaking. The few hardy companies that survived the '20's ran headlong into the Great Depression, but all managed to survive till mid century. Characteristically, these consisted of four dollar-watch producers and three conventional watch manufacturers (see Figure 2). They made millions more watches after 1930 but life was tenuous and the heyday of American watchmaking was long gone.

Clearly, a lot of things happened to American watchmaking between 1850 and 1930, as it was swept by trends in the economy, society, and fashion. Table I shows changes in watch styles as the industry progressed. Data is given for demarcation dates of the four industry phases, plus the century date, at which time dollar-watch manufacture matched jeweled watches in quantity and ladies' pocket watches reached peak production.

Watch sizes progressively diminished. The industry began in 1850 producing solely 18 sizes, then smaller movements each had their period of industry dominance until wristwatches took over during the 1920's. Production of 16-size watches peaked in 1910, along with production of highly jeweled movements for the railroad trade. The slim 12-size watch gained maximum popularity around 1920, then declined as wristwatches took over.

Seven-jewel watches were the staple of the industry, losing out to fifteen-jewel movements only as wristwatches became popular. Sales of watches bearing nine to thirteen jewels was a factor solely in the early days of the industry since many European watches were so jeweled. Movements of seventeen jewels or more did not become widely produced till railroad watch standards were issued after 1892. The railroad watch market saturated in 1910, after which manufacture of highly jeweled, as well as 16-size, movements fell off.

Production rates for the jeweled watch industry maintained a steady rise till peaking in 1910 at a level of 2½ million watches per year. By then the American watch industry was rapidly weakening and production declined to Nineteenth-Century levels before the Great Depression arrived. Dollar watch production, which began around 1880, had caught up with jeweled watchmaking by the turn of the century, continuing a rapid rise to seven million watches per year till depression gripped the country in 1929.

While the beginning of this century marked the start of decline for United States watch manufacture, it must be noted that American watchmaking made several technical innovations in recent years, continuing to promote change in a tradition bound industry. The first was introduction of electric watches in the early 1950's. These operated on the same principle as oscillating displays seen in grocery stores, providing a reliable replacement for the self-winding mechanism. Second came solid-state electronics in two distinct steps during the early 1960's. One was solid-state switching for electric watches, replacing mechanical switches in earlier models. The other was introduction of tuning fork timekeepers, which required electronic control circuits as well as complex mechanical devices. Finally, American industry produced the completely solid-state watch during the 1970's, using a quartz oscillator. Traditional foreign makers greeted these developments as merely interesting fads; they failed to perceive electronic timekeeping emerging as the future trend, and are now suffering turmoil as watch sales return to American soil. Some of these developments were initiated in familiar watch companies of old, but those companies have since passed away and electronic watchmaking is full of new names.

It is lamentable to see the remnants of our watchmaking history disappear, but change is bound to continue. It is easy to feel comfortable with past traditions and fear unfamiliar new ideas. There must have been similar feelings a century ago when American watch factories created massive changes. They pulled the time honored watchmaker away from his bench, but his exit was not in vain. The new order bestowed benefits upon millions of people and those "new" watches are now friendly heirlooms.

Table I
Watch Trends — 1860-1930

Year	Movement Size % Production					Jewels % Production				Yearly Production Millions	
	18	16	12	L*	W**	7	9-13	15	17+	Jeweled	Dollar
1860	(100)	—	—	—	—	(65)	17	18	—	0.01	—
1880	75	4	—	18	—	65	—	35	—	0.5	0.2
1900	35	15	7	(40)	—	60	—	25	15	2.0	2.0
1910	34	(29)	(10)	20	6	51	—	20	(29)	(2.5)	5.0
1930	—	7	8	5	(80)	42	—	(40)	18	1.0	(7.0)

Note: Circled figures are the largest in their respective columns.
*Ladies' watches
**Wristwatches

1. The Time Honored Watchmaker — Abraham Louis Perrelet — Invented the Self-winding Watch ca. 1775.

SECTION 1

FORMATION OF THE INDUSTRY

I
BEFORE THE BEGINNING

The time-honored watchmaker had been toiling at his bench since the invention of watches in 1500.[1] Peter Henlein (1480-1542) of Nurenburg, Germany, supposedly made the first watch about then, but at the same time a more recognizable version emerged in Southern Europe, probably Italy. By 1575 the German design was dying away while Italian-style watchmaking had spread to France, Switzerland, and England. A watch was more a development than an invention, being merely a smaller version of spring-driven table clocks that had existed for at least a half century. While earliest models were the size of an apple, they were a feat of miniaturization considered remarkable at the time. Their mechanism consisted of two brass plates and a balance cock housing a verge escapement, mainspring with fusee, and a train of three brass wheels. This design remained amazingly unchanged for more than 300 years and formed the basis from which all developments sprang till 1800. With almost perpetual persistence, its last vestiges could be seen in gilded fullplate movements made by the American industry right into the 20th Century.

Improvements in watchmaking came slowly. Early watches were extremely poor timekeepers till 1675 when the balance spring, or hairspring, was invented in England. Even with the balance spring progress was slow, and another century passed before precision timekeeping was achieved. A few other advancements were being made, for by 1700 the English were beginning to use wheel cutting engines. These eliminated hand-filed gears and the finer trains were much more capable of regular performance, one of the earliest hints that machinery could improve watchmaking. Also the English were taking steps to subdivide the labor of watchmaking among specialists. Thomas Tompion was among the first to organize a large shop of divided labor. The result was much more efficient production, which accounted for Tompion's large output of over 5000 watches. It was probably for this reason that Tompion invented serial numbering, to keep matched parts together as they progressed around the shop.

Horological progress accelerated through the 1700's. Escapements were invented in England and France which permitted better timekeeping, most importantly the cylinder, duplex, lever, and chronometer. English makers also developed the art of jeweling, allowing consistent wear-free performance for a long period. The greatest impetus to precision watchmaking was the Longitude Act, passed by the British Parliament in 1714. There was no practical means for determining longitude, or east/west location, which was a serious problem for the seafaring English. The Longitude Act provided up to £20,000 reward for any satisfactory method of solving this dilemma. Though it took a few years, money proved a strong incentive and in 1759 John Harrison unveiled his fourth attempt at a navigational timekeeper, amounting to a large watch with temperature compensation on the balance spring. (The timekeeper solution is based on the east/west time shift we recognize today with time zones.) Harrison's #4 performed well enough to qualify for the £20,000 and precision watchmaking was on its way. Close behind came London makers John Arnold and Thomas Earnshaw to perfect detent (chronometer) escapements, adjusting of isochronous balance springs, and bimetallic balance wheels for temperature compensation. Moreover they refined these to simple forms that could be manufactured by skillful watchmakers

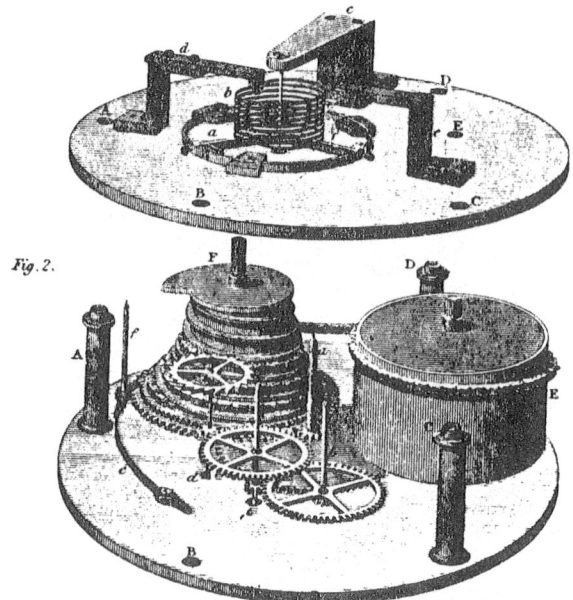

Fig. 2.

2. English marine chronometer ca. 1790 — showing fusee and train.

in sufficient quantity to serve the English fleets. With development of chronometers, England was world master of precision watchmaking. Detent escapements with temperature compensated balances and isochronous helical hairsprings were used in the finest English watches. Cylinder and duplex escapements sufficed for high-grade watches not intended for navigation, and some were as finely executed as chronometer watches. But verge watches still constituted the majority of pieces since they were durable and simpler to make. Furthermore, accuracy of a quarter hour a day was adequate for most needs, which the verge could deliver. The English developed variations in style among these types of watches but all were still based on the gilded fullplate movement of antiquity with fusee and chain. Specialization had resulted in a growing network of shops that supplied particular items like dials, hands, fusee chains and verges. In addition, considerable business had built up in Lancashire and Prescot, near Liverpool, supplying materials and tools to the English trade. A major part of that business was in ebauches, i.e. rough movements complete with rough trains but no escapement or dial, which were the starting point in making hand-crafted watches. These were made in considerable quantity on special machinery and loosely conforming to standard guidelines. Still, there was no automation, no finished parts were made, and virtually nothing was interchangeable.

Precision watchmaking was not being ignored on the Continent, but the French and Swiss were behind the English. French makers were producing some chronometers after 1760, but their most important developments were for customers who didn't require such accuracy. It was more important that common timekeepers be easy to carry and stylish enough to please the customer's fancy. Beside decoration, the most important feature was to achieve a thinner watch, and the fusee was the biggest problem in that regard. In curious contrast, fusees were required for the precision of chronometers, and in lowly verges since the verge was very sensitive to changes in motive force. The possibility of eliminating the cumbersome fusee presented itself in frictional rest escapement such as the cylinder,

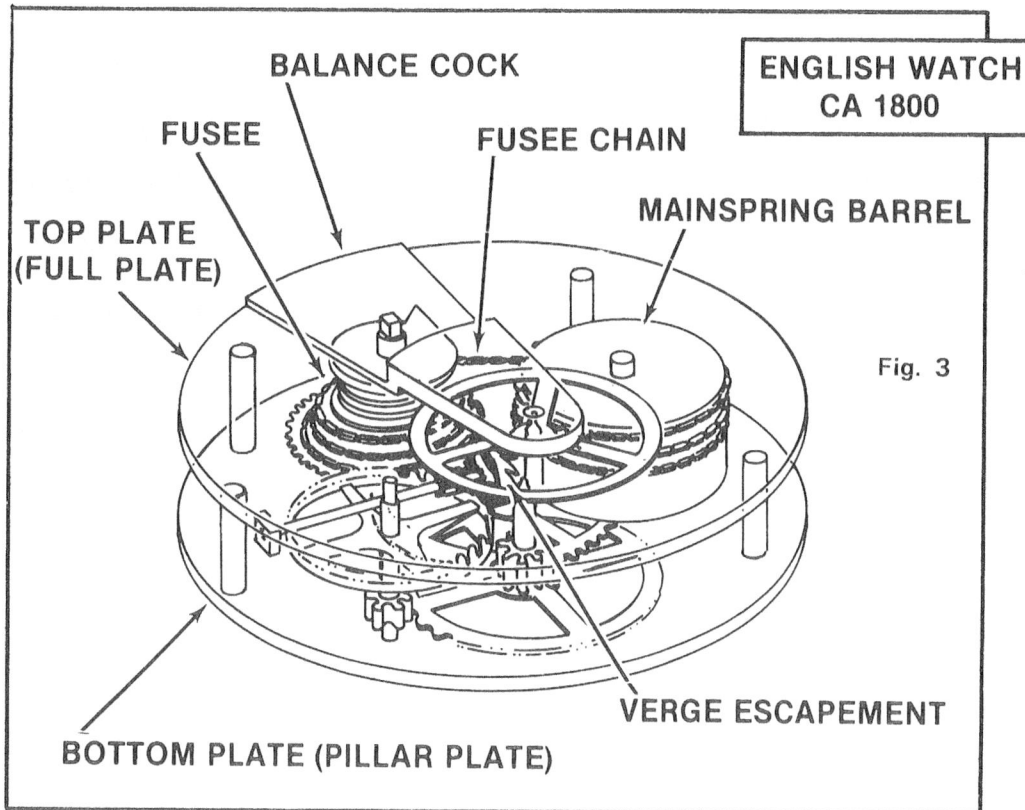

ENGLISH WATCH
CA 1800

BALANCE COCK

FUSEE

FUSEE CHAIN

MAINSPRING BARREL

TOP PLATE
(FULL PLATE)

Fig. 3

VERGE ESCAPEMENT

BOTTOM PLATE (PILLAR PLATE)

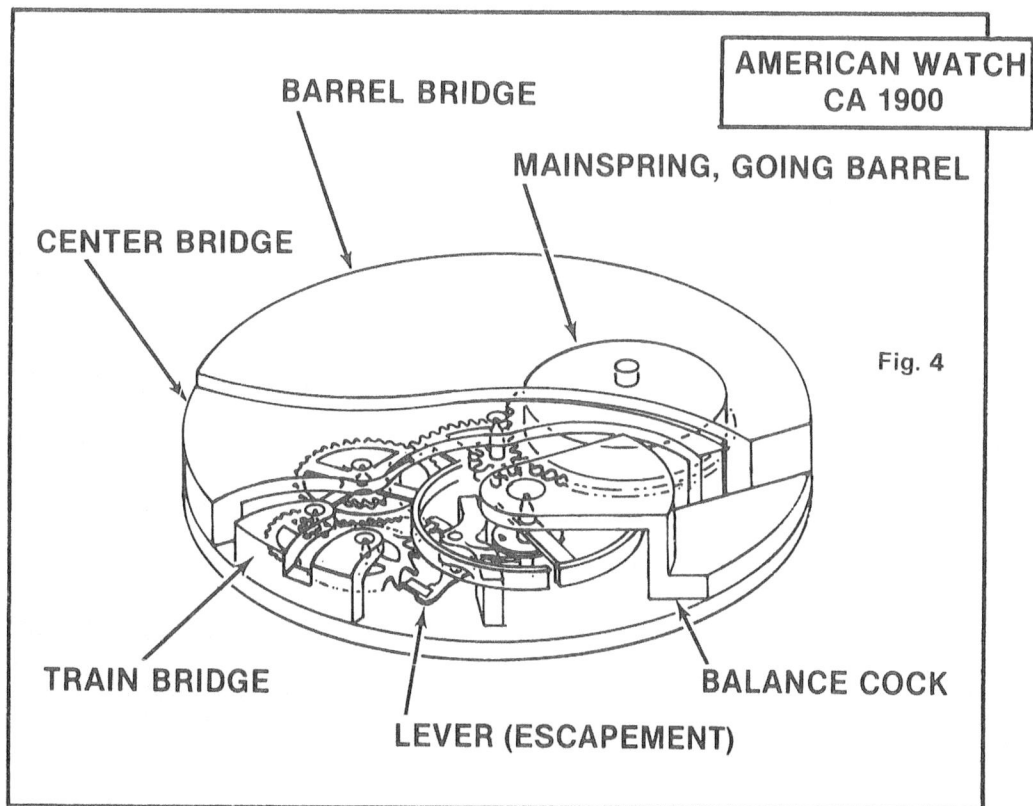

AMERICAN WATCH
CA 1900

BARREL BRIDGE

MAINSPRING, GOING BARREL

CENTER BRIDGE

Fig. 4

TRAIN BRIDGE

LEVER (ESCAPEMENT)

BALANCE COCK

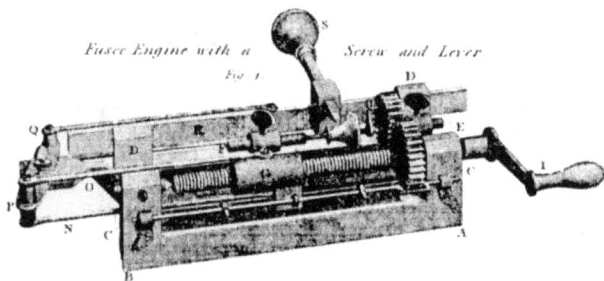

3. English fusee cutting engine ca. 1770 — one of many types.

4. Swiss or French fullplate verge watch ca. 1800.

duplex and virgule, which tended to compensate themselves for variations in driving torque and were also adequate timekeepers. Antoine Lepine of Paris finally abandoned fullplate construction in the 1770's, supporting the wheels under separate bridges. This allowed the edge of the movement to be thinner than the middle, so along with a going barrel to replace the fusee and frictional rest escapement, Lepine calibre watches became quite slim and sleek. Abraham Breguet of Paris began building his fabulous career during the 1780's with major emphasis on modern appearing watches of Lepine calibre. He combined style with superior craftsmanship to produce watches that were both fashionable jewelry and excellent timekeepers. His overwhelming reputation created a new world of watchmaking design, starting the traditional "turnip" down the road to oblivion.

Up to 1750 Swiss watchmaking had been concentrated around Geneva as part of the jewelry trade, for Geneva was a commercial city with gold and silversmiths exporting all over Europe. As demand for watches grew, between 1750 and 1800, watchmaking spread to the mountain and lake region along the French border north of Geneva. This became a cottage craft to which people turned during winter when farming was impossible. Since they were interested in an export commodity, they restricted themselves to moderately priced watches that were not demanding to make. There was more emphasis on appearance than function and the common product was a typical verge and fusee in the old French style, with a pretty case. Then, as Paris fashion brought forth thin watches, Swiss makers turned to Lepine-style bridge watches with going barrel. Frictional rest escapements were more complicated to make, but this expense was balanced by eliminating the fusee. The overall product was simpler, more attractive to many shoppers, and better suited to the purposes of the Swiss industry.

As horology progressed into the 19th Century, French watchmaking steadily declined and was not a major influence after Breguet. However, Breguet's influence was monumental. Popularity of his fashionable watches was becoming universal, in addition to which he spread the use of cylinder escapements and his own form of club tooth lever. Swiss watchmaking was growing from cottage craft to full scale manufacture, concentrating on Lepine calibre movements with cylinder and lever escapements. Slim Swiss watches, mostly of mediocre quality, were being exported to countries around the world, and by 1850 serious Swiss craftsmen had also developed sophisticated stemwind watches with high grade levers and excellent adjusting. As with the English they used considerable division of labor and had shops specializing in specific parts as well as tools and ebauche making. In fact, Frederic Japy had developed a number of specialized tools as early as 1780 for

quantity manufacture of Swiss ebauches. To some degree the Swiss continued to advance watchmaking machinery, more than in England where traditional methods prevailed. At mid-century, Swiss watchmaking was extensively developed and its finest makers had worldwide reputations.

The British retained their watchmaking reputation during the 19th Century and a large segment of their craft remained dedicated to a legacy of high quality. English pocket chronometers were the finest watches made anywhere while duplex escapements remained their second level of quality till halfway through the century. Cylinder and verge watches had rapidly died away at the beginning of the century when rack levers enjoyed a brief period of popularity. Through the second quarter of the century the English form of detached lever was blossoming and by 1850 replaced all but a limited segment of duplex and chronometer production. English watchmakers retained the fusee but with Breguet's influence on fashion, thick fullplate watches gave way to ¾ plate movements, some of which

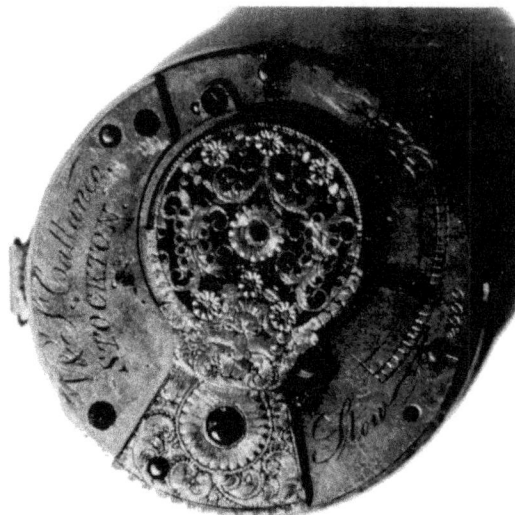

5. English fullplate verge watch ca. 1800.

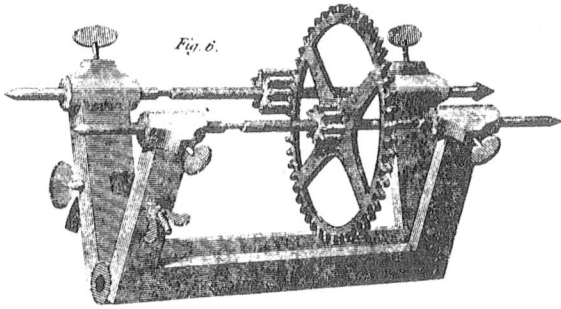

6. Depthing tool — for determining proper center-to-center distance between wheels.

were quite small and thin, fusee and all. Division of labor was directed as much toward decreasing cost as increasing quantity and the excellence of English specialists remained unequalled. In 1841 Edward Dent estimated that a London-made watch, including the case and counting the fusee chain as one piece, comprised 166 parts made by 43 different workmen.[2] He also pointed out that one great advantage of the lever escapement, which was just then gaining full popularity, was that it lent itself to greater subdivision of labor than other escapements.[3] And while the 1854 directory of Prescot, England, listed only five watch and clockmakers, there were 120 ebauche makers, since the area supplied ebauches to London and Liverpool, as well as 20 balance makers, 18 escapement makers, 31 pinion makers, 28 wheel makers, and 43 tool makers.[4] Except perhaps for the ebauche makers, London would have had a similar array of individuals. The maker whose name appeared on the watch had to manage its evolution through this maze of specialized shops till a finished timekeeper emerged. Watch frames were obtained from ebauche makers, then carried about to workers in a day when there was no rapid transportation. Though often handled in groups, movements sometimes traveled one at a time, to the finisher, the engraver, the gilder and so on.[5] A large shop might have employed 20 or more workmen but more usually work was done by individuals in shops adjoining their homes. Consequently much of the English craft revolved around independent individuals who could not afford complicated machinery, which would simply have eliminated their jobs. Nor did they care for the idea of lowering the quality of English watchmaking, for machinery could not have produced the exquisite work they did by hand. Therefore the English handcraft system continued to make a limited number of fine watches by traditional methods for the traditional customer. As a result, the Swiss were rapidly capturing lower priced and stylish parts of the market, even in London stores, and the English were left with a dwin-

dling portion of the high priced market. America had been an important export market for the English and with the rise of the American industry they suffered a considerable loss. In addition, Swiss and American industries continued to improve the quality and reputation of their best watches, further injuring the English craft. Britain faced the mechanized competition too late, never adequately industrializing their watchmaking, and by 1900 had become fatally crippled.

The history of American horology does not have quite the ancient background of European watchcraft. Also, little colonial watchmaking was practiced here compared to early American clockmaking. The 1760's probably saw a few individuals making watches in America and homemade watches may have cropped up in several of the colonies about then. These early watchmakers were likely English colonists, plus a few Dutch and French, who were trained in Europe and brought their craft with them. While our watchmaking history perhaps began about that time and in such a fashion, virtually no contemporary watches or written accounts survive to complete the details. About the earliest lodged claim to distinction was the 1807 obituary of Thomas Harland of Norwich, Connecticut, which specifically stated he was the author of the first watches made here, but this certainly does not constitute historical proof. Harland came to the colonies in 1773 and billed himself as a watchmaker during the Revolution, but later advertised as selling imported watches.[6] Like many of those who advertised watchmaking skills at the time, he was a London trained Englishman who worked here on English made tools brought over with him. In fact, for most watches that may have been made here during the period, many of the parts, like the tools, would have been imported from English specialty shops. Whatever the details, the major influences surrounding the early American watches would have been English skill, tools and parts.

Early makers in America had no choice but to use roughly the same methods as their English counterparts, but with fewer people involved. The first well documented watchmaker was Luther Goddard (1762-1842) of Shrewsbury, Massachusetts, who descended from an old New England family that came from England in 1665.[7] His father married the sister of Benjamin Willard of Grafton, whose four sons became clockmakers, and at the age of 16 Luther apprenticed to his cousin Simon Willard at Grafton. He later returned to Shrewsbury, supporting himself by farming during the summer and clockmaking during the winter in a small building on the farm. When the Jefferson Embargo cut off imports in 1809 Goddard began making watches with his sons Parley and Daniel plus some apprentices. In addition he apparently hired several Englishmen who were watchmakers by training but had come to America as soldiers during the Revolution. They made tradi-

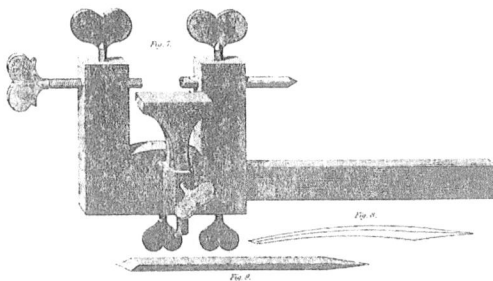

7. Drilling lathe and spade drills — lathe to be turned by bow — one of many specialized lathes used by watchmakers.

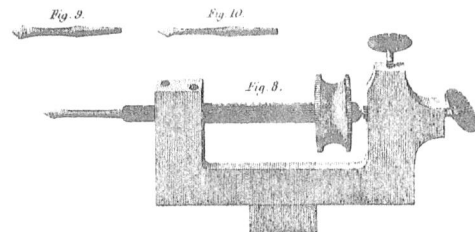

8. Large turns and gravers — part to be turned was mounted between stationary centers, turned by bow, and cut with hand held graver (see illustration 1).

9

Table II
NUTSHELL HISTORY OF WATCHES

NOTE · MECHANICAL TIMEKEEPING ORIGINATED APPROXIMATELY 1300 AS WEIGHT DRIVEN TOWER CLOCKS WITH VERGE ESCAPEMENTS.

1400

─ SPRING DRIVEN TABLE CLOCK

AVERAGE DAILY ERROR

1500 ─ WATCH-SPRING DRIVEN CLOCK SMALL ENOUGH TO CARRY

- VERGE, FUSEE   (ITALY
- FULLPLATE   GERMANY
- ONE HAND   FRANCE)

2 HOURS ─ WATCHMAKING IN ENGLAND AND SWITZERLAND

1600

─ POCKET WATCHES

30 MIN.   WHEEL CUTTING ENGINE ─ ─ BALANCE SPRING

DIVISION OF LABOR ─ ─ MINUTE HAND

1700

15 MIN. ─ WATCH JEWELS

ALL MAJOR * ESCAPEMENTS

─ TEMPERATURE COMPENSATION

10 MIN.   ISOCHRONOUS HAIRSPRING ─

EBAUCHE MANUFACTURE ─ ─ BRIDGE CONSTRUCTION

─ SECOND HAND

1800

5 MIN.

─ LEVER ESCAPEMENT PERFECTED

AMERICAN INDUSTRY BEGINS ─

30 SEC.   DOLLAR WATCH ─ ─ METALS INSENSITIVE TO MAGNETISM AND TEMPERATURE

1900

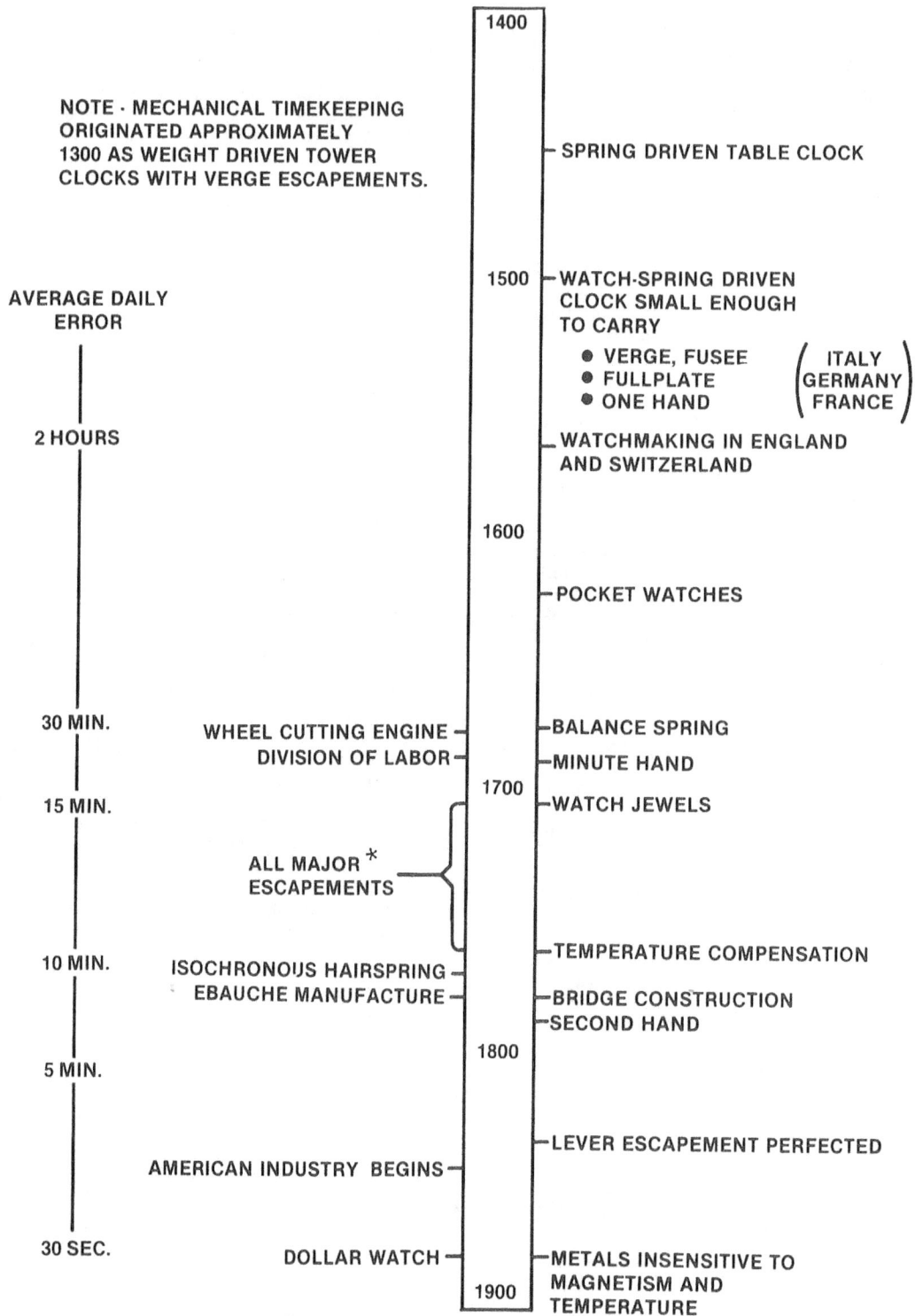

*DEBAUFRE, RACK LEVER, CYLINDER VIRGULE, DUPLEX, LEVER, DETENT

ENGLISH FULLPLATE

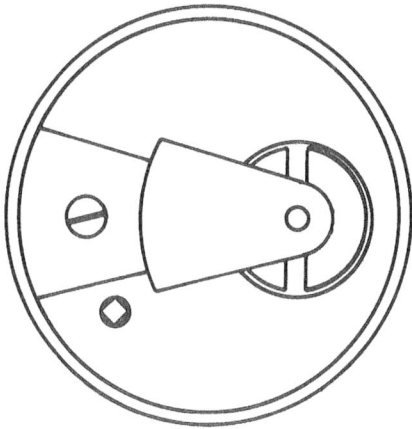

ENGLISH ¾ PLATE

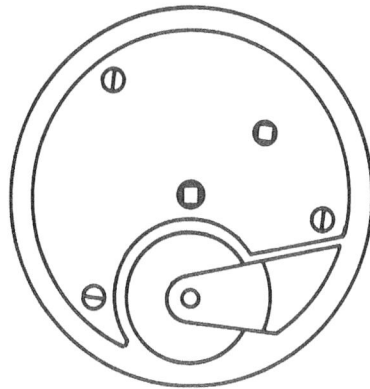

Fig. 5
MOVEMENT STYLES

FRENCH BRIDGE STYLE

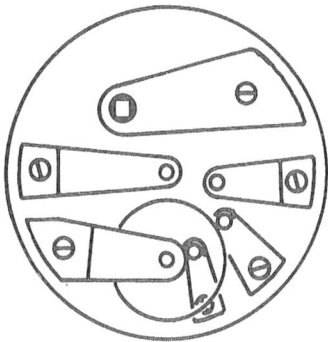

SWISS BRIDGE STYLE

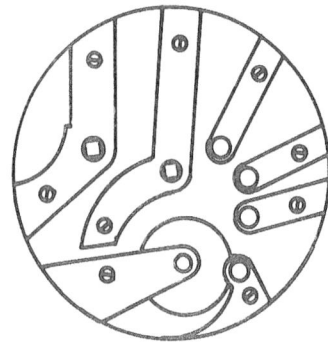

9. Luther Goddard watch #235 — marked L. Goddard.

10. Luther Goddard watch #462 — marked L. Goddard & Son Shrewsbury.

tional English-style watches of good quality with fusee and verge escapement and commanding a fair price for the day. Hands, dials, mainsprings, hairsprings, verges, steel pinion wire and chains were bought from material dealers in Boston who imported them from England. Otherwise, all the plates, wheels and brass parts were cast in Shrewsbury, the watches and cases were finished in Goddard's shop and engraved locally. About 530 watches were made this way till production ceased in 1817, and Luther's son, Parley, finished approximately 60 more during the next few years. Goddard's watchmaking was brought to a close by resumption of trade after the War of 1812. He could not compete with the prices of imported watches, especially the Swiss, and was forced out of business. In making his watches he used only the common watchmaking tools of the day, all imported from England. Thus there was little unusual about Goddard's watchmaking except that it was done in the United States.

Two apprentices who worked for Luther Goddard were Jubal Howe and William Keith. Jubal Howe continued in the watch business and was foreman in the Boston jewelry shop of Jones, Lows and Ball when they hired a young man named Aaron Dennison, who later began factory watch production in America. William Keith joined Dennison's Boston Watch Company, played a role in the company's move to Waltham, and became president of the American Watch Company of Waltham in 1861. This intertwining of people and events was common in the history of American watchmaking and was certainly a major cause of the extreme uniformity that developed within the industry. The writings left by Mr. Keith shed light on the watchmaking of Luther Goddard and also point out that others were making watches during the period. Jubal Howe made a few at Medway, Massachusetts. Another apprentice of Goddard's named James Harrison made perhaps 80, and a Mr. Wheelock of Sutton, Massachusetts, made watches using some variation of the Debaufre escapement, which had a brief period of popularity in England about 1790. Keith also reported a rack lever watch made by Julien Montandon in New York around 1835[8] but Montandon was an importer and the rack lever was probably English.[9] Another

person who definitely did make watches was Jacob Custer (1810-1879) of Norristown, Pennsylvania, who made about a dozen watches of his own design between 1840 and 1845. So while Luther Goddard became one of the best known of early American watchmakers and was probably the most prolific, he was by no means alone in plying his craft. These other makers were much like Goddard, using traditional methods to make their watches; parts were hand-made one at a time at the watchmaker's bench.

By Goddard's time the days of traditional watchmaking were rapidly becoming numbered. In fact, the influences that led to an American watch industry in the 19th Century had begun to gain strength about the time of the American Revolution. It is speculated that Thomas Harland may have built some watches in that period, but he definitely made a large number of clocks, more than any of his contemporaries. He had a mind for efficient production and in this vein expressed ideas that it would be advantageous to introduce interchangeable parts into clockmaking. His foremost apprentice, Daniel Burnap (1759-1838), was also a prolific maker who actually did use some part standardization in his clocks; but it was in turn Burnap's apprentice, Eli Terry, who brought these ideas to fruition. In 1806 Terry purchased a gristmill in Plymouth, Connecticut, where he set waterpower turning crude clockmaking machinery. There he constructed 4000 grandfather clock movements in three years using simple jigs and fixtures, and mass production became a reality in clockmaking. To assist in this job he hired, among others, two carpenters named Seth Thomas and Silas Hoadly, so at that point Thomas Harland's ideas had reached their most ambitious advocates.[10] Eli Terry did not stop there; he eventually developed the Connecticut shelf clock, using machinery to make both the movements and cases, and by 1840 the wooden works shelf clock sold for as little as $8.

Eli Terry was among several Americans who began using machinery for mass production. In New Haven, Connecticut, Eli (cotton gin) Whitney had received a government contract to build 10,000 muskets in two years, starting in 1798. He did not manage to complete the order till 1809 but

12

11. At the Bench — Nineteenth Century Swiss watchmaker with bench by windows. Hand driven flywheels drive countershafts on the bench. Countershafts drive lathes and drilling machine (in vise). A selection of bows is on the wall behind the workman.

13

during the decade of effort loudly promoted his "uniformity system" for producing interchangeable parts. He may not have achieved interchangeability at all and the first government contract to specify interchangeable parts was written in 1813 for Simeon North of Middletown, Connecticut, to deliver 20,000 pistols.[11] The growing American arms industry continued to promote mass production on increasingly sophisticated machinery. Early tools required constant attention from an operator to oversee them and possibly supply the power. Some later machines became semi-automatic, performing their function and shutting themselves off, once loaded and started. The most important thing was that machinery allowed workmen to repeat an operation in rapid succession, keeping all parts sufficiently identical as to be interchangeable. Hand fitting was thereby eliminated and production increased. These techniques were so effective that by 1850 America was exporting both guns and gunmaking machinery to England, France, Russia and other countries.

So, very early in the 19th Century the Connecticut area had two home-grown models of large-scale manufacture with interchangeable parts: North's guns and Terry's clocks. Therefore it was not too surprising that the first attempt to make watches by similar methods also occurred in Connecticut. That was the venture of the Pitkin brothers, Henry (1811-1846) and James (1812-1870) of East Hartford. The Pitkins were an established New England family that had come to America from London in 1659, settled in the Hartford area and owned land that now comprises the center of East Hartford. Through the years members of the family had undertaken various types of manufacture including woolens, cotton, paper, glass, cloth, and gunpowder during the Revolution. The watchmaking branch of the family was the 6th generation in America, being the youngest of four brothers who started a business in silver flatware and operated a successful jewelry store in Hartford. The three oldest brothers had learned watchmaking and silversmithing from a local jeweler named Jacob Sargent, whereas the youngest, James, learned the crafts in the family shops and was really more of a businessman. While the business concentrated on silversmithing, Henry, the second youngest, was primarily a watchmaker and mechanical inventor, conceiving during the 1830's what he named the "American Lever Watch." He and James constructed a model watch as well as models of some machines Henry designed to make interchangeable parts. When they judged that all looked well, the family erected a new building in East Hartford and took in apprentices to begin making machines for the real effort. There were delays, since the building was also used for silversmithing, but in 1838 the watch was ready for sale, opening a new age in watch manufacture.[12]

The Pitkins intended to make the complete watch in their own shop but had to settle for imported dials, hands, mainsprings, hairsprings, and balance jewels. They manufactured the rest, punching plates and wheels from rolled brass, fabricating lantern pinions (one of the rare appearances of lantern pinions in watches), making balances, levers and pallet jewels, and hiring John R. Proud to make the cases.[13] The design was completely original, eliminating the fusee common to English watchmaking, but incorporating the form of lever escapement which the English had begun to make respectable in high grade watches. As with English levers, the escape wheel had pointed teeth, which was unusual in the later development of American watches. Also uncommon was their early attempt to use conical pivots on the arbors, running in hardened screws set in the plates, as on the balance wheels of today's cheap alarm clocks. But by far the most unusual feature compared to

THE PITKIN WATCH.

12. Pitkin watch #66 — from an early cut.

any watch up to that time was the use of machinery to make interchangeable finished parts. They were by no means completely successful in achieving interchangeability, nor would they be the last to discover that such a goal was difficult to reach. Yet, they were the first to progress so far, meeting with moderate success. About 350 watches were sold through the jewelry store in Hartford, and as far as is known today these were sound timepieces that pleased their owners for years. Seeking good fortune, Henry and James set out in 1841 for New York City, center of the jewelry trade, where they hoped to operate on an expanded scale. The established trade did not readily accept their novel watch over the cheaper Swiss imports, so they desisted from watchmaking in 1842 and it is doubtful that more than 500 were made in New York. Innovative and restless, Henry grew increasingly resentful that the world could turn its back on such a work of genius and died in 1846 after a complete mental collapse. James continued with a jewelry and watchcase manufacturing business in New York but never made another attempt to produce watches. To the Pitkins must go credit for originating machine manufacture of interchangeable watch parts, at least in America. With their sophistication they were able to produce more watches than Goddard in half the time. Their efforts did not give rise to an industry but that was certainly not because of any deficiency on the part of their watch or manufacturing methods. Had they remained in Hartford where their family business was known, they might have established a sound company, enticing other entrepreneurs to follow. Nothing succeeds like success and success had to await more fortunate pioneers.

Mass production was not being ignored by the rest of the world. One of the most advanced of early systems for producing interchangeable watch parts was that of Pierre Frederic Ingold (1787-1878), a Swiss watchmaker who had done jewel work for the renowned Breguet.[14] As innovative and restless as Henry Pitkin, he developed machinery for making watchplates, balances and wheels between 1825 and 1835, then attempted to introduce his system in France, England and New York with little success. His ideas were ingenious and efficient for the period, for instance produc-

14

ing a complete watchplate in one lathe setup. The British Watch and Clockmaking Company was probably his closest brush with good fortune, resulting in a small watch output between 1842 and 1845, but resistance of the craft guilds stifled this. He always managed to subscribe for his ventures endorsement from enlightened scientists and watchmakers, but could never steer them to success. Unfortunately, little is known of his sojourn in New York during the 1845-1855 period. He reportedly claimed that a later American company used machinery he left there,[15] but no remnants have surfaced that would support this, nor did American watchmakers use Ingold's techniques. He developed machine capability equal or superior to the Pitkins' and likewise might have given birth to an industry had he designed and marketed his watch in suitable fashion.

Several other Europeans made considerable progress with machine production at the same time the Pitkins were trying it in America. George Leschot (1800-1884) was another Swiss watchmaker who, by the 1830's, had begun to concentrate on using machinery for making lever escapements. These efforts qualified him to join Vacheron and Constantin in 1839 where the company was endeavoring to standardize their movement styles. With backing from the company, Leschot developed tools for ebauche making as well as mechanical manufacture of various other watch parts. This capability enabled Vacheron and Constantin to produce more movements than they could use, and they sold the excess to other Geneva firms.[16] The newly formed firm of Patek Philippe & Cie. also endeavored to use mechanical methods as much as possible during the 1840's. These companies tended to keep their methods secret so that, like the Pitkins, they failed to trigger the rise of an industry.

This is unfortunate, for only by "catching on" and spreading can an idea become fully useful. Historians of European watchcraft strive to point out through these and other examples that Americans cannot take the entire credit for inventing modern watch manufacture. This is obviously true; the impact of the industrial revolution was being felt on both sides of the Atlantic. One result was that both Europeans and Americans were on the verge of machine watch manufacture utilizing new ideas of mechanical production. Ideas are useless unless they are followed by action. The important point is that European watchmakers as a whole refused to embrace the industrial philosophy till Americans threatened to bury them with it. For whatever reasons, Americans turned philosophy into large scale action, developing an industry that changed watchmaking the world over.

It can be seen that a watchmaking industry was not to spring suddenly from a vacuum. Numerous advances had already been made. Machinery was used extensively in areas such as ebauche making and several companies had tried to make finished watches by machine. The industrial trend was rising in fits and starts within the ancient craft. But starting an industry required more than an inventive

stroke, for invention abounded and one can only wonder at how long it was ignored. It remained for sufficient ingenuity to combine with sufficient business skill in an atmosphere that could accept innovation. Nineteenth Century America was an ideal time and place for such developments, Nationalistic pride promoted "Yankee Ingenuity" and favored homemade products, even over superior imports. So America was where industrial watchmaking was born.

REFERENCES

1. A general background of European watch history may be obtained from several sources. Among them:
   a) Baillie, Clutton & Ilbert, *Britten's Old Clocks and Watches and Their Makers* (New York, New York, Bonanza Books, 1956);
   b) Cedric Jagger, *The World's Great Clocks and Watches* (London, Hamlyn, 1977).
2. George C. Eckhardt, *United States Clock and Watch Patents 1790-1890* (New York, 1960), p. 65.
3. Edward Dent, *On The Construction and Management of Chronometers, Watches and Clocks* (reprinted Exeter, NH, Adams Brown Co.), p. 16.
4. Alan Smith, quoted by Kenneth Roberts in *The Lancashire Watch Company* (Fitzwilliam, NH, Kenneth Roberts Publishing Co., 1973), p. 13.
5. Ibid., p. 12.
6. Chris Bailey, *Two Hundred Years of American Clocks & Watches* (Englewood Cliffs, NJ, Prentice-Hall, 1975), p. 65.
7. Dr. P. L. Small, *Luther Goddard and His Watches* (BULLETIN, National Association of Watch and Clock Collectors, April 1953), p. 355.
8. Dr. P. L. Small, *Luther Goddard and His Watches Part III* (BULLETIN, National Association of Watch and Clock Collectors, April 1954), p. 181.
9. Jaquet & Chapuis, *Technique and History of the Swiss Watch* (Olten, Switzerland, 1953), p. 147.
10. Bailey, p. 104.
11. Edwin A. Battison, *Muskets to Mass Production* (The American Precision Museum, 1976), p. 10.
12. Dr. P. L. Small, *The Pitkin Brothers* (BULLETIN, National Association of Watch and Clock Collectors, October, 1954), p. 251.
13. Charles S. Crossman, *The Complete History of Watchmaking In America* (reprinted Exeter, NH, Adams Brown Co.), p. 4.
14. Carrington and Carrington, *Pierre Frederic Ingold and the British Watch and Clockmaking Company* (Journal of the Antiquarian Horology Society, London).
15. T. P. Camerer Cuss, *The Country Life Book of Watches* (Middlesex, England, Hamlyn Publishing Group Ltd., 1967), p. 101.
16. Jaquet & Chapuis, p. 176.

II

THE BEGINNING

The founder of the American watch industry was Aaron Lufkin Dennison. Dennison (1812-1895) was an inventive mechanic, adventurous businessman, apostle of the idustrial revolution, and salesman of contageous enthusiasm. His active mind ranged over a variety of business ventures but watchmaking was his training and the area in which he hoped to excel as a mechanical pioneer. The Dennison family had come to America in the late 1690's when George Dennison was engaged aboard a British man-of-war by press gang. He left the navy in an equally informal manner, eventually establishing himself in Annisquam, Massachusetts, as a ship owner[1] and store keeper.[2] Upon his death, several sons removed to land he had owned in Freeport, Maine, where the family multiplied and employed themselves in craft and commerce. Aaron was born a shoemaker's son at Freeport, the fifth generation of Dennisons on American soil, but seeking more prosperous opportunities, the family moved to Brunswick, Maine, and Aaron was apprenticed to a local clockmaker. At age 21 he journeyed down to Boston in order to learn more of watchmaking and try his hand at a repairing business. He was soon in the employ of Jones, Lows & Ball under their shop foreman, Jubal Howe. Howe had been an apprentice in Luther Goddard's watchmaking shop at Shrewsbury, Massachusetts, in the early part of the century, and Dennison must have been entranced by stories of that first significant watchmaking effort in America. Aaron next went to New York where he learned more of watchmaking skills from European craftsmen at work there, but in 1839 he was

back in Boston. This time he established himself in a watch tool and material business, and it was in organizing his materials that he developed the Dennison Standard Gauge, still used to measure mainsprings.[3] He always remained in touch with his family and through this period supplied both technical and financial support to sundry family business schemes, such as silk farming and his brother Eliphalet's unsuccessful foray into jewelry retailing. One thing he found was that his material business was accumulating a sizable line of small cards, tags and boxes for the jewelry trade, all imported from France, which he thought could be made quite easily. Consequently he contrived some simple machinery to facilitate their manufacture and arranged for the family to take up the business on the farm in Brunswick. This served the father well till his retirement and prospered under management from Eliphalet, eventually becoming the Dennison Mfg. Company, now of Framingham, Massachusetts.[4] Such ventures also kept Aaron laboring in debt, maintaining himself through a talent to persuade creditors. By the late 1840's Aaron's own business had expanded and he hired Nelson P. Stratton to help at watch repair. If he hadn't heard of the Pitkins' machine-made watch before, he did then, since Stratton had been one of the four apprentices who helped build the Pitkin machinery in 1837. (In fact, one obituary listed Stratton's middle name as Pitkin.[5]) In the meantime Stratton had worked for a while at the Springfield Armory, where rifles were being produced on an interchangeable part basis, and spent some time at watch repairing. Whatever the source of inspiration, Dennison began to seriously consider the idea of manufacturing watches by machinery. He made visits to the Springfield Armory, built a cardboard model

AARON L. DENNISON.

13. Aaron L. Dennison.

EDWARD HOWARD.

14. Edward Howard.

of a factory and spent many evenings strolling on the Boston Common pondering how to approach the task. Apparently he had had ideas of this sort as far back as his apprentice days in Maine,[6] and the time spent later with Jubal Howe certainly would have left him dreaming of them again. Undoubtedly Stratton was also very helpful in this period since he had been part of the only previous such attempt in America. Dennison was the sort who turned ideas into action, and in 1849 sought out Edward Howard with the idea of starting a watch factory.

At that time, Edward Howard (1813-1904) was already established and successful. He had been born in Hingham, Massachusetts, and at age 16 was apprenticed to Aaron Willard, Jr., in the Boston suburb of Roxbury. After leaving Willard he joined with Henry Plimpton making fine bank scales. Then in 1842 he formed a partnership with fellow apprentice David P. Davis and Luther Stephenson to make a general line of balances, gallery clocks and Willard-style timepieces (banjos). As a businessman Howard proved to be clever as well as ambitious. Since a federal law had been passed setting postage rates by weight, he entered five different designs in competition for a government contract to make balances, winning an order for 40,000 machines. Next he designed a system of weights, measures and a balance, all housed in one cabinet and conforming to a standard system of measure established by the Massachusetts Legislature. He then obtained a contract to supply one of these complete sets to each city, town, and county in Massachusetts, totalling 330. At various times he had also made fire engines, sewing machines, and leather splitting machinery, so it can be seen that like Dennison, Howard had an organized mind plus an ambition to join in enterprise.[7] It has been said that upon hearing Dennison's plans of machine manufacture he suggested they go into business making steam fire engines, but Dennison persuaded him that watches were the key to the future.[8] Whether or not this really happened, the two did reach an agreement in 1849 to form a watchmaking company.

In typical fashion, Dennison provided ideas while convincing others to supply money, spinning neverending projections of productivity and profit. (A later associate said that Dennison "could make more figures showing results than any man I ever met with."[9]) Howard and Davis produced a combined sum of $10,000 while the real support for this scheme was Howard's father-in-law, Samuel Curtis.[10] Curtis was an integral part of the Roxbury clockmaking community, a cousin to Aaron Willard, Jr., and brother of Lemuel Curtis who developed the girandole wall clock. Samuel began as a dial painter, developed skills as a gilder and reverse painter, and amassed a steady fortune as a mirror maker.[11] His financial support was at least as instrumental as Dennison's ingenuity in developing factory watch manufacture. He maintained complete faith in the plans of Dennison and Howard, providing up to $80,000 over the next few years.[12] This fertilized the growth of the American watch, for which Curtis received little return or recognition, and he only withheld further funds out of self preservation.

With company and capital organized, Dennison withdrew from his other business interests and went to England while Howard outfitted a place in his Roxbury clock factory for work to commence.[13] In England Dennison observed the manner of British watchmaking but was primarily concerned with locating sources of materials he would be needing. After returning to America in 1850, he began to experiment on designs for a watch, as well as machinery to make it, but just starting in the field of mechanical watchmaking, a number of these designs were not satisfactory. In fact, Dennison turned out to excel more in the overall task of superintendent than as hardware designer.

CHARLES S. MOSELEY.

15. Charles Moseley.

Firing the Dials.

16. Firing dials.

17

His idea for a timepiece was in the style of an English watch since Americans preferred English to Swiss imports. He was struck with the notion of an eight-day watch, feeling this would be a logical convenience, for eight-day clocks had replaced the one-day timekeeper in the home. Unfortunately, eight days' running was difficult to achieve in a pocket watch, requiring very careful design and construction. Dennison's design, about 18 size with one going barrel, was not working well and in early 1852 Nelson Stratton rejoined Dennison at the factory with a plan for changing over the movement to a one-day timekeeper, utilizing material on hand since a number of parts had been prepared. Even as work continued on Stratton's one-day design, Dennison began again on an eight-day model. This time he had David and Oliver Marsh design one with twin mainsprings, and work commenced to build 100 of these, though few were completed. In the line of machinery, Dennison had built an upright plate turning lathe and a tool for punching all holes in the plate at one time. These did not operate so well in production and the machinery probably progressed more under the efforts of Charles Moseley, who joined the operation in 1852, and began to develop what became the American factory lathe. Clearly, Dennison was not able to supply all the brainpower required, but did not hesitate to bring in people who could.

As a result, Stratton and Moseley were followed by a long list of people who joined the young company in Roxbury. The Marsh brothers, who had worked in Howard's clock factory, handled plate making and production of trains. James L. Baker was hired as a general machinist, but was soon turned to making screws and levers, punching levers out then filing them by hand, stacked in a fixture. Traditional balances were made in steel and gold by an English balance maker brought to this country for the job, and a Scotchman made the first dials. Because of problems with the dials, John T. Gold was brought into the company, son of a New York dial maker. The first watch hands were imported but a person was soon recruited to make them, and cases were made by a casemaker from Providence, Rhode Island. Yet another man by the name of John Lynch came from New York to do jeweling. In those early days, Mr. Lynch set the bottom jewels all flush with the pillar plate. He then laboriously determined how deep or shallow to set the top jewels by putting each wheel between the plates and noting the height of the arbor. At any rate, the family was growing, and so was their home, for in 1851 the group occupied a new building outfitted with a steam engine across the street from Howard's clock factory.

Numerous problems plagued the early efforts to build watches at Roxbury. Since dial making presented continuing difficulties, John Gold was finally sent to Liverpool, England, to learn how dial making was practiced there, and Edward Howard carried out experiments with dials for many years after. Pallet jewels were found to be wearing escape wheel teeth because they were finishd from top to bottom, presenting a burnisher-like finish to the teeth. Finishing jewels from heel to toe eliminated the problem. Pointed tooth escape wheels were also found to be somewhat damage prone due to the delicate tips and were eventually changed to the club tooth form used by the Swiss, though more than 5000 pointed tooth wheels were used. As their first watches neared completion the group discovered that they could not produce a good gilded finish on the plates. So Nelson Stratton went to Coventry, England, to learn gilding, though English methods were secret and Stratton had to use "diplomatic" means to obtain information. As with Dennison's first machinery, it was also found that some machines and methods they hoped to use would not work satisfactorily. For instance they tried cutting wheels stacked on a mandrel, but for lack of alignment the stack came out tapered, so an English wheel cutting engine was employed till the stacking technique was perfected. Some parts they didn't attempt to make at all; jewels were imported as well as pinion wire and probably hairsprings and mainsprings. Pinion wire came from England in one-foot lengths which had been drawn through a series of dies to form rough pinion leaves. These were then cut to the length required for individual arbors, the ends finished, and the leaves in the remaining pinion section finished and polished. Despite all the problems, success was at hand. Specialized lathes were developed for turning arbors and plates, with all their required recesses and cavities. To a high degree, steel parts were punched, then finished or shaped on power-driven gang files.

The result of all this effort was that, in late 1852, 17 of the eight-day watches designed by the Marsh brothers were presented to investors under the name "Howard Davis & Dennison," and in the spring of 1853 100 30-hour watches went out marked "Warren." For a brief period in 1851 the company had taken the name "American Horologue Co." but since they were obliged to obtain material from England, where the competition would not be well received, the name was changed to the innocuous "Warren Mfg. Co." This was in tribute to General Joseph Warren, who had been born near the site of the Roxbury factory in 1741 and killed in the battle of Bunker Hill in 1775. When watches began to leave the factory, the company confidently adopted the name "Boston Watch Co." Following the 100 "Warren" watches were almost 900 marked "Samuel Curtis," after the chief investor in the enterprise, and factory watch production was a reality in America.

These sold for about $40 in their cases, a substantial price in their day. The movements were gilded, fullplate models of about 18 size, basically what is known today as the model "57" Waltham. They had the general appearance of an English watch, especially with large aquamarine jewels, similar to those in some Liverpool watches, yet they embodied numerous features which became standard to American watchmaking. The most fundamental difference from English practice was the use of a going barrel rather than fusee and chain; this was a major simplification and the fusee was never incorporated in an American factory watch. Also the movements were held in the case with

17. Warren Mfg. Company's first watch — 8-day Howard Davis & Dennison, 1852.

screws rather than hinged out the front, as in England. In this period Dennison, with his penchant for organizing, began to use his standard series for sizing. This was based on the English Lancashire gauge, the size number of the movement being the number of thirtieths of an inch by which the pillar plate exceeded 1-5/30 of an inch. An 18 size watch was thus 1-23/30 inches in diameter, although many of the early watches did not exactly fit this scheme. The balances of early movements were plain rather than temperature compensated, with hairsprings secured to the top plate and geared to beat 16,200 beats per hour. This was faster than the 14,400 beats per hour in many English watches, but the "quick train" of 18,000 beats eventually became most common in American watches since it made a more stable timekeeper. All in all, it cannot be said that these watches reached new heights of horological excellence, but more important, they made an adequate success of themselves and established what became the basic American watch for many years.

After 1853 the Boston Watch Company got busily into its career. With increased production, the new factory soon became too small, and the cost of additional in-town land was high. Dennison began considering land along the Charles River in Waltham, and after discussions with Mr. William Keith became interested in the farm on which Keith had previously lived. Through formation of a land company, with W. Keith as treasurer, the farm was purchased from the current owner, Mr. James Brown of Little Brown & Co. Publishers in Boston, who became a shareholder in the land company. Construction began on a new factory and, in keeping with Aaron Dennison's desire to promote technical innovations, this was the first concrete factory in America. It was occupied in October of 1854, the Boston Watch Co. then having about 100 employees struggling to produce 10 watches per day. Usually more like 6 were made, but it was only through Dennison's and Howard's many connections in the jewelry trade that such a sizable output could be sold. These ranged from 7 to 15 jewels and were marked C. T. Parker, Dennison Howard & Davis, or P. S. Bartlett. The Boston watch was slowly finding acceptance in the trade, promoting gradual growth at Waltham till the country began experiencing a business recession in 1856. The firm may have already had difficulties with earnings due to Dennison's tendency to experiment rather than concentrate on the boring task of production. The general business decline precipitated a complete crisis. Under an informal working arrangement, Dennison had been managing the shop while Howard tended company finances. With disaster looming, action was necessary, so Dennison took to the road to sell his contagious brand of enthusiasm to new creditors. Considerable sums were borrowed and Fellows & Schell, a New York jewelry wholesaler, pledged $20,000 in return for becoming agent for the factory's output. (There were perhaps 100 movements made marked Fellows & Schell.) In addition, Charles Rice, a Bos-

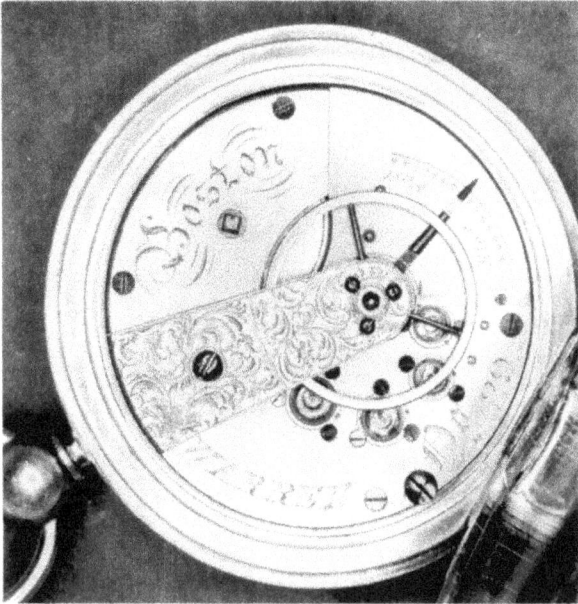

18. Warren #29 — 1853 — Boston Watch Co. Courtesy of George Townsend.

19. Samuel Curtis #396 — Boston Watch Co., Roxbury, Mass. 1853 or 1854.

The Waltham Factory in 1857.

20. Waltham factory of Boston Watch Co. 1857.

21. Dennison Howard & Davis #3074 — Boston Watch Co. ca. 1855.

22. Tracy Baker & Co. #5012 — Tracy Baker & Co. 1857.

23. P. S. Bartlett #9123 — Appleton Tracy & Co. 1859.

ton shoe merchant, was granted a mortgage on watch material and movable property in the factory. Sales dropped through the winter of 1856 while debts climbed, so that bankruptcy became inevitable, and the Boston Watch Co. was sold at auction in April of 1857.

Considerable intrigue and scheming transpired during the months before the auction. Charles Rice was a friend of Edward Howard and the two planned for Rice to obtain the factory and contents on the day of the auction. Dennison evidently did not relish the thought and scurried forth on a plan of his own. It's difficult to estimate how many parties he attempted to interest in purchasing the plant. He got the most response from Mr. Baker of Tracy & Baker, Philadelphia makers of gold watch cases to whom the Boston Watch Co. owed at least $8000. Baker caught the fire from Dennison, and Tracy finally tumbled, on the proviso that a third party be found to furnish a majority of needed capital. Tracy and Baker finally found an interested party in Robbins & Appleton, New York watch wholesalers (Fellows & Schell declined to become involved). Both D. F. Appleton and Royal Robbins saw a future in the watch factory and Robbins was retiring from wholesaling for health reasons. Dennison estimated the plant would auction for $30,000 to $35,000 plus a $10,000 mortgage, and Robbins agreed to enter the bidding as a ⅔ partner if Tracy and Baker would take the other ⅓. On the day of the auction everyone arrived separately, with Mr. Robbins authorized to bid at his discretion. Charles Rice bid heavily for the factory, but when the gavel dropped the Waltham plant belonged to Royal Robbins for $56,000.[14]

24. Plate making department at the factory. Benches are along the windows, where watchmaking had been practiced for centuries.

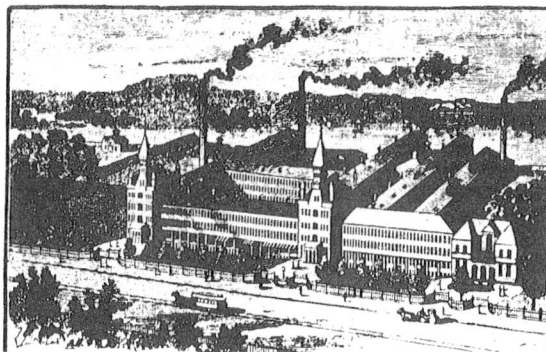

25. American Watch Co., Waltham 1880.

The fact that people fought to purchase the Boston Watch Co. indicated that Dennison and Howard had succeeded technically, even if they fizzled financially. The American watch was being accepted by both the jewelry trade and the buying public. Acceptance in the trade was essential, for jewelers were the link between the factory and consumers. Dennison had courted them carefully. His standard sizes meant that with a modest number of interchangeable movements and cases, jewelers could offer customers a large variety of combinations, which foreign watchmakers could not match. Jewelers could invest in less inventory, an obvious benefit, but of no value if customers could not be found. So early movements were styled after English watches that the public preferred. Dennison had one other thing to popularize his watches among the jewelry trade: spare parts. Since his factory could supply interchangeable finished material, he could save jewelers the task of custom-making repair parts, which European watches required. As confidence in American watches grew, and lower prices made them more attractive, they took on distinctive features and gained their own identity. Eventually, the Swiss copied American styles, even to standard sizes. Such was hardly the case in 1857, but Robbins knew the watch business, and knew he had bought a good thing.

As majority owner, Robbins fully intended to take charge of the business he bought, but his style was to do so from the background and the new organization was named Tracy Baker & Company. Baker proved a bit fickle, rapidly selling out to Robbins, so after only a month the company name was changed to Appleton Tracy & Company. Appleton was Mr. James W. Appleton, older brother of Robbins' wholesaling partner, who represented the new owner at the factory. The greatest problem facing the new company was to survive the recession that felled the Boston Watch Company. A cheaper model was designed, but otherwise no new expenditures were made. Even so, workers accepted a 50% pay cut to avoid temporarily closing the plant, and production was held to a minimum. Waltham endured, the recession abated, and Robbins gradually built the Waltham enterprise up to mammoth proportions. Robbins owned all but Tracy's 15%, which Tracy sold to another party after about 7 years.[15] The company name was changed to the American Watch Co. in 1859 and again to the American Waltham Watch Co. in 1885, but company management remained Royal E. Robbins till his death in 1902.

The factory superintendent remained Aaron Dennison till his discharge in 1862. Robbins was a merchant and businessman, not a mechanic, so that Dennison was indispensable as the new company got on its feet. On the other hand, Dennison was pronounced in his ideals concerning the watch business and could not refrain from voicing his opinion on company direction. But Dennison was only a salaried employee, no matter who invented American watchmaking, and the management did not enjoy such a powerful dissenter on the staff. As the country heated toward civil war in 1862, Dennison began agitating for yet a cheaper model which, he understood from contacts in the trade, would sell well among soldiers. But the company was still recovering from the recession, felt the management, so it was not time to finance a new product. Dennison continued that opportunity was wasting, and the directors removed him from the company.[16] Dennison was effective in organizing the money and support of others, but it naturally followed that the others owned the organization. It happened in the Boston Watch Company, the American Watch Company, and would happen to Dennison again.

When Royal Robbins purchased the factory in 1857, the contents did not include everything that had been the old Boston Watch Company. Under his chattel mortgage Charles Rice removed watch material and machinery, apparently more than he was entitled, and returned it to the old factory in Roxbury. For the next year Edward Howard, followed by about 15 Waltham workmen, acted as manager there in the interests of Mr. Rice, finishing old Boston Watch Co. material under the name Howard & Rice. By 1858 Howard was able to settle with his creditors and purchase the company outright, combining with his nephew, Albert Howard, as E. Howard & Company. Under this new management watches began appearing, characteristic of Howard's desire to advance American watchmaking. These were a sort of half-plate watch, the earliest having plain balances, that became known as the "seven pillar" model. They were approximately 18 size with the first 18,000-beat train in an American watch, Reed's patented mainspring barrel, and soon bimetallic balances for temperature compensation. Along with good adjusting, these features resulted in the finest pocket timepiece yet produced in an American factory.[17]

George P. Reed, who came with Howard from the Boston Watch Company, had patented a stationary mainspring barrel to protect the gear train in the event the mainspring broke. No such precaution was necessary with a fusee and chain, but the breaking mainspring would uncoil rapidly back out against the inside of a going barrel, subjecting the barrel teeth and center pinion to an immense impact which often resulted in damage. Reed's barrel was the first device in an American watch to protect the train from this shock, and by no means represented Reed's last contribution to horology. The stationary barrel was set solidly in the pillar plate, with the outer end of the mainspring hooked to it and the inner end of the mainspring attached to a

26. Appleton Tracy & Co. grade #502,722 — American Watch Co. ca. 1872.

27. Howard & Rice — Boston (Roxbury) ca. 1858. The barrel bridge still retains its Boston Watch Co. marking and the BWC serial number. Some Howard & Rice do not have the BWC marking, but all retain the BWC serial number sequence.

wheel meshing with the center pinion. If the mainspring broke it merely sprang out against the stationary barrel with no damage done.

Howard thought this safeguard was a worthwhile addition and likewise began introducing other features which he hoped would advance factory watchmaking. Some were experiments which did not last, such as Mershon's rack lever regulator, Cole's resilient escapement, upright pallets, and helical hairsprings. Others were more permanent, like Reed's whiplash regulator which became popular throughout watchmaking. In 1865 Howard began making his own tempered hairsprings and around 1870 introduced stem-winding and nickel finish. He developed his own series of movement sizes, starting with "A" for 1" in diameter and progressing up the alphabet for each 1/16 of an inch. His first watches were "N," slightly larger than an 18 size, and about 1871 he came out with an "L" which was just smaller than a 16 size. After Howard left, the company also adopted a unique manner of designating adjustments to the balance and hairspring. A hound stamped on the movement meant it was unadjusted, a horse meant adjusted for isochronism and temperature, a deer was for isochronism, temperature and position, and these were the first American watches adjusted to six positions. Howard never sacrificed quality to achieve quantity, so that the total production after more than 40 years was barely above 100,000 watches. His efforts established a reputation for building superior watches commanding superior prices. This maintained a strong company till 1903 when it was sold to the Keystone Watchcase Company, who introduced new high-grade models bearing the name E. Howard Co.

One other pioneering effort began after the Boston Watch Co. collapsed. This was the Nashua Watch Company, which was short lived, but like Edward Howard, endeavored to produce a high-grade watch in the United States. Even more ambitious than Howard, the group at Nashua strived to introduce greater machine production and interchange-ability than so far had been achieved. While Howard wanted to leap ahead of the mediocre quality of the Boston Watch Company, he felt that a certain amount of hand finishing was required to do this. Even some of the automatic machines he used were designed to fit one piece uniquely to another, creating "married parts" which were not interchangeable.[18] One goal of the Nashua Watch Co. was to design tools and equipment that would make high-grade parts accurately enough to be completely inter-

E. HOWARD & CO'S

28. The Howard factory in Roxbury — original location of the Boston Watch Co.

29. Howard #425 — early movement with plain balance.

31. Howard #2879 — with Mershon's patent regulator ca. 1860.

30. Howard #438 — similar movement to #425 (illustration #29) having temperature compensated balance wheel.

32. Howard #11231 — typical N size Howard key winder ca. 1862.

33. Howard #219,304 — typical N size Howard stem winder ca. 1880's. Reeds whiplash regulator and deer, indicating that the movement was adjusted for temperature, isochronism, and position.

35. Nashua #1227 — Nashua Watch Co. ca. 1862. English ¾ plate style.

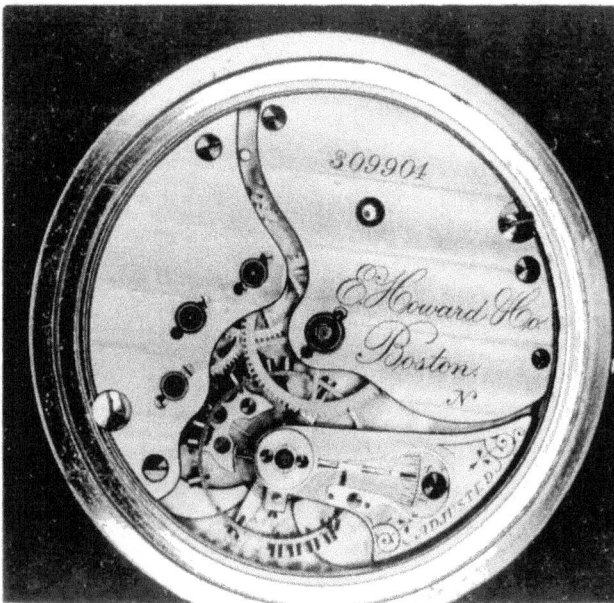

34. Howard #309,904 — late Nineteenth Century stem winder, 17 jewels.

36. Appleton Tracy & Co. grade #100,914 — American Watch Co. ca. 1863. 20-size key-wind movement based on the Nashua design.

changeable. The founder of the Nashua Watch Co. was Belding Dart Bingham, a jeweler in Nashua, New Hampshire, and maker of fine clocks. He had gone to work at Waltham in the late 1850's to learn something of watch manufacturing, and while there became friends with Nelson Stratton. The two became convinced they could manufacture a precision watch, so they went to Nashua in 1859 and assembled a small factory, taking from Waltham Ira Blake, Charles Vander Woerd, Charles Moseley, James H. Gerry, James Gooding, and others. The basic watch had been designed by Mr. Bingham after fine English ¾-plate movements of the period. The English had begun discontinuing fullplate construction about 1840 in favor of the thinner ¾-plate style. Notable Swiss makers were also building sound reputations around thin bridge model watches. So like Howard, Nashua followed this trend away from the thick fullplate watch. They used a lever escapement and added Stratton's mainspring barrel, which differed in detail from Reed's stationary barrel, but had the same purpose of protecting the watch from a broken mainspring.

When the model watch was complete, all parts were carefully measured and machinery prepared to duplicate them.[19] It was later reported that three dial foot holes in the pillar plate were used as a master tooling datum. The various machines and tools each had three precision pins to locate the plate for every operation. In retrospect it may be ventured that holes for the three pillars were the likely datum, allowing matching holes in both pillar and top plates. This master datum was then used for machining recesses in the plates, laying out the train, drilling pivot holes, and setting jewels, all without depthing tools or measuring individual parts. Wheels were machined in stacks of 50 using a fly cutter, while pinions were made on a cam-operated lathe which accurately reproduced tooth profiles and shoulder-to-shoulder arbor lengths. With precision arbors, jewels could be set into the plates to a uniform depth while still maintaining proper end shake for the arbors. Charles Moseley was the master mechanic and Charles Vander Woerd was his assistant, designing at that time his machine for accurately forming the cutters to make gearteeth. Clock and watchmakers had long known that ideal wheel teeth have an epicycloidal shape, but they could only form cutters by eye to achieve the best approximation, making good wheel cutting specialists prized and necessary for precision watchmaking. Charles V. Woerd's invention generated mathematically correct shapes on the cutters, allowing superior trains, and was probably one of the first approaches to geartooth generation in America. From here he began a career as a fine mechanic, designing machinery for the American Watch Co. and becoming instrumental in the formation of the Waltham Watch Tool Co. and the United States Watch Co. of Waltham.[20] Also at Nashua, Mr. Gooding developed the capsule method of making temperature compensated balances, which became standard in the industry. The traditional method of making bimetallic balances was to melt brass onto the steel core of the balance. In the new method a steel blank was punched and turned to a precision diameter. A close fitting ring of high brass was pressed onto the steel core and the pair placed into a tight fitting capsule of low brass. This whole package was then fused in a furnace without solder, the capsule remaining separate from the brass balance rim due to its higher melting temperature. From there the bimetallic wheel could be finished, machined, the crescents punched, and the rim drilled and tapped to receive balance screws.[21]

The Nashua company accumulated a work force of about 35 people and material for 1000 watches, both 16 and 20 size. In the spring of 1862 they had run out of money, and

with the country engaged in a civil war, additional capital was difficult to find. Therefore, after brief negotiations, they were absorbed by the American Watch Co. of Waltham and established there as a completely separate department, with their own tool shop. The superior Nashua machinery was a technical advance for Waltham, and the fine design of the Nashua watch gave Waltham a high-grade product to compete with E. Howard and quality imports. This department continued to develop the ¾-plate watch, specifically the Waltham models 62, 68, and 72, and also the fullplate model 70, which began a line of movements long used in the railroad trade. Associated with all these were advances in stemwinding, finish, and quality that served to raise standards throughout the industry, making ¾-plate style the preferred design for American precision watches till 1900.[22] The highest-grade movements of the Nashua department, in combination with the production capability of their machinery, were part of the Waltham display at the Philadelphia Centennial Exposition which shocked the Swiss industry into adopting American manufacturing methods.

Beside creating machinery, the pioneers had introduced the three types of watch movements that would dominate the industry in years to come. English style fullplate watches were manufactured at Waltham, English style ¾-plate movements were designed at Nashua, and Howard used a split-plate design of his own. Fullplate construction was a major influence for a half century since it was the most economical to manufacture. After being absorbed by Waltham, Nashua's ¾-plate design became an important factor for 40 years, being widely used throughout the industry. Howard's type of split-plate design was eventually displaced by a more Swiss version at most companies, but its fundamental advantage of easy assembly remained unchanged. After 1900, split-plate construction was universal.

By 1860 the American watch industry was soundly established. The Boston Watch Co. had pioneered machine watchmaking. Through Dennison's efforts, techniques were developed to make a moderate quality watch on tools that would produce interchangeable parts. That was actually the easy job, for there were many mechanics capable of designing the necessary machines, and some of it had been done before. Dennison had the initiative and ability to pull together a company, find support in the jewelry trade, and put forth a product acceptable to the trade and its customers. Though his initial objectives were a bit optimistic, he quickly learned not to underestimate the massive task, or over estimate the requirements of the product. People just wanted a decent watch, nothing more, and by 1857 the Boston Watch Co. could supply one for under $20. It cost them $150,000 to learn how. As rain poured over the bankruptcy auction, it must have been easy to dismiss the Boston watch as a dismal misadventure. Considerable money and effort had been consumed to produce an undistinguished watch. Only a few visionaries inside the business could see the potential that had become available. As far as Royal Robbins was concerned, the pioneering was over. He had a golden goose that would manufacture plenty of watches when the economy improved and was ready for them. He was a practical businessman, confident that practical mechanics could provide whatever the public wanted.

Idealistic watchmakers were slightly less satisfied. As far as Howard and Stratton were concerned, the watch was mediocre, and machines were still crude and inefficient. Everything had to be improved. They raised the quality of their watches, bringing a new level of credibility to American factory timepieces. Much more importantly they increased the capability of their machinery, especially at Nashua. Boston Watch Co. technology could not have pro-

25

The Watch Factory at Waltham, Mass.

37. American Watch Co. at Waltham ca. 1870.

duced enough watches of any quality to supply the vast American markets that were to open in the near future. Millions of people would be satisfied with garden variety watches if millions of them could be made. High quality watches had their place too, and railroads would be demanding them soon. But the demand would grow to large numbers and only sophisticated factory methods could fill such an order. The task of the pioneers, beyond managing to survive, had been to develop watchmaking capability, not watches. Technical advances made by Dennison and his associates had assured that factory watchmaking would remain a sound business.

REFERENCES

1. Charles W. Moore, *Timing A Century* (Harvard University Press, Cambridge, MA, 1945), p. 4.
2. M. Copeland & E. Rogers, *The Saga of Cape Ann* (Bond Wheelwright Co., Freeport, ME, 1960), p. 172.
3. Charles S. Crossman, *The Complete History of Watchmaking in America* (reprinted Adams Brown Co., Exeter, NH), p. 12.
4. Moore, p. 7.
5. Dodge & Lucchina, *Jonas G. Hall* (BULLETIN, National Association of Watch and Clock Collectors, Inc., October 1976), p. 437.
6. Moore, p. 5.
7. Crossman, p. 13.
8. Henry G. Abbott, *A Pioneer* (reprinted Adams Brown Co., Exeter, NH, 1968), p. 35.
9. E. Tracy to C. N. Thorpe (reprinted BULLETIN, National Association of Watch and Clock Collectors, Inc., April 1949), p. 603.
10. Moore, p. 11.
11. Dr. P. L. Small, *Samuel Curtis* (BULLETIN, National Association of Watch and Clock Collectors, Inc., December 1952), p. 283.
12. Tracy to Thorpe, p. 604.
13. Generally, the technical background of Dennison's Boston Watch Company has been extracted from Crossman.
14. Tracy to Thorpe, p. 604.
15. Ibid.
16. Moore, p. 44.
17. Crossman, p. 55.
18, Henry G. Abbott, *Watch Factories of America* (reprinted Adams Brown Co., Exeter, NH), p. 23.
19. Frederick Mudge Selchow, *Belding Dart Bingham — The Nashua Watch Co.* (BULLETIN, National Association of Watch and Clock Collectors, Inc., December 1975), p. 555.
20. Thomas De Fazio, *The Nashua Venture and The American Watch Company* (BULLETIN, National Association of Watch and Clock Collectors, Inc., December 1975), p. 581.
21. Crossman, p. 208.
22. De Fazio, p. 585.

III
COMPANIES OF 1864

Failure of the Nashua Watch Co. in 1862 left only two factories making watches in the United States: the American Watch Co. of Waltham and E. Howard & Co. in Boston. (Charles Fasoldt, then in Rome, New York, was the only other notable source of American watches.) Both were struggling to gain strength after rising out of the old Boston Watch Co. and for the moment the picture was hardly one to attract new investors. Fortunately, the Civil War provided impetus to retail watch sales, with wholesalers and importers enjoying considerable volume. Much of this probably rested with the fact that a watch was one of the few personal items that a soldier could carry, otherwise it was attributable to general prosperity in industrial communities during the war. Actually, E. Howard & Co. muddled through the period with poor to indifferent prospects. This was due to Mr. Howard's intent to produce high quality rather than quantity, which priced him above the mainstream of the market. Up the Charles River at Waltham, where the ideals and price tag were more modest, sales during the war years were excellent and the American Watch Co. experienced one of the most prosperous growth periods in its long history. The result was regular cash dividends, which never fail to attract businessmen. So, in 1864, with the war still raging, five new watch companies were formed:

1. National Watch Co. of Elgin, Illinois
2. Newark Watch Co. of Newark, New Jersey
3. United States Watch Co. of Marion, New Jersey (Jersey City)

J. C. ADAMS.
38. John C. Adams

4. Tremont Watch Co. of Boston, Massachusetts
5. New York Watch Co. of Providence, Rhode Island

All five organizations had people from the jewelry trade as their founders or financial backers. While it can hardly be said that all the companies had long and successful careers, they did demonstrate the character of the emerging industry. With confirmed confidence in the venture that Dennison and Howard started, the watch business came to life with a rapid rate of birth, and death. New entries to the field did not intend to start from nothing, the way Dennison did. They sought their knowledge and manpower from Waltham, Howard or one of the later existing factories. One could thus see the same people appearing at factory after watch factory, with the necessary result that American watchmaking became highly uniform in style.

Incorporation of the National Watch Co. in Elgin, Illinois, was an excellent example of new company formation; new and old faces appeared, investors were found, and machinery began to accumulate. In the early 1860's Nelson Stratton had discussed the possibility of forming a western watch company with John C. Adams of Chicago, but Stratton went to England for the American Watch Co. and the matter was dropped. In the spring of 1864, Mr. Adams was visited by Ira Blake (formerly with Stratton at the Nashua venture) and P. S. Bartlett, both of American, and the issue was revived. Some sort of promising agreement was reached for Adams then went casting about for investors and business minds to make a company work.

This sort of activity proved to become a habit with Adams, who was influential in organizing at least five watch companies: Elgin, Cornell, Illinois, Adams & Perry, and Peoria. He was born in Preble, New York, but the family moved to Elgin, Illinois, in 1842 where he served a five-year apprenticeship to an old Liverpool watchmaker. After various watchmaking jobs he became manager of the watch department for the Chicago firm of W. H. & C. Miller in 1861, and the next year became timekeeper for the railroads around Chicago. He rightly felt that by the 1860's Chicago was a good location for a watch company. Chicago had started as a tiny trading village on the river and lake transportation routes long used by Indians and explorers. It was still a small village well into the 1820's, but as a strategic trade location, it grew as the west grew. Chicago proved equally strategic to the railroads, so that the 1850's saw explosive growth of a city to which industrial goods flowed from the east and from which farm products were shipped from the west. In this thriving atmosphere Adams found much of the backing he sought in the person of B. W. Raymond, former Chicago mayor and promoter of the railroads. For a location he settled on his hometown of Elgin, situated on the Galina & Chicago Union Railroad, which had been the earliest line into Chicago. The fortunes of Elgin had been improving since trains first chugged into the little community in 1849. It already had a successful condensed milk plant operated by Gail Borden, whom Adams attracted as an investor, and the businessmen of Elgin donated a factory site to draw J. C. Adams' little watch venture.[1]

By the end of 1864 the organization was in place, machine tools had been purchased in the east, a temporary shop erected, and it was time to bring in people to commence work. The method of manning the operation was to recruit from the most obvious place. Offering a contract for high wages, land to homestead, and the opportunity to see their own company grow, the Adams group lured from the American Watch Co. P. S. Bartlett, Charles Moseley, Otis Hoyt, George Hunter, D. R. Hartwell, C. E. Mason, D. G. Currier, and J. K. Bigelow. These were all prime, knowledgable individuals and their mass exodus was a se-

THE ELGIN WATCH FACTORY.

READY FOR THE CASE.

READY FOR THE POCKET.

39. National Watch Co. — Elgin, Illinois, ca. 1870.

40. Elgin B. W. Raymond — 15-jewel railroad grade ca. 1870.

vere setback for the Waltham plant. (Elgin was no more pleased when half of them were snatched away to the Illinois Springfield Watch Co. after 1869. Also an interesting sidelight, the Elgin factory was designed by H. H. Hartwell, brother of D. R. Hartwell, a Boston architect who later designed watch factory buildings for Waltham and Waterbury.) Their presence completed the list of ingredients and in the beginning of 1865 the National Watch Co. of Elgin began making machinery for watch production. This was a long and difficult operation to carry out properly, with no watches being completed till 1867. (Only one of the companies of 1864 got out watches before 1867.) The cost of supporting such an effort for years with no sales revenue was one of the stumbling blocks many companies never overcame, and was solely a function of how well the organizers did their planning. The first watches were 18 size, gilded, fullplate, keywinders designed by Charles Moseley and Daniel G. Currier, similar in appearance to movements made at Waltham. These were named the B. W. Raymond model and due to their high quality were soon selling well in the railroad trade. Business continued going well for the young company and over the years they expanded in an organized manner, bringing out new models and continually designing new equipment to keep their factory up to date. Because of this aggressive and well managed policy, only Elgin surpassed Waltham in production of jeweled watches.

Such prosperity was not common to all watch companies. A brief look at the other companies of 1864 brings to view the twins of New Jersey: Newark Watch Co. and the United States Watch Co. of Marion. Both were formed by New York watch and jewelry firms, got modest starts in

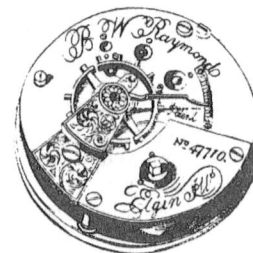

1863 by setting up their machine shops, then moved to factories staffed with men from the old Boston Watch Company. Also, both had difficult careers and failed after a few years.

The Newark Watch Co. was formed by Louis S. Fellows and Robert Schell of Maiden Lane, New York, who had paid A. L. Dennison $20,000 to become sole agents for the Boston Watch Co. just prior to the latter company going bankrupt.[2] They got a shop started in New York with the help of Napoleon B. Sherwood, who had been with the Boston Watch Co. and also designed machinery for E. Howard. In 1866 Fellows died, Sherwood left New York, possibly after a disagreement, and their watchmaking shop moved to an old hat factory in Newark, with Arthur Wadsworth its superintendent. The operation was named Robert Schell & Co. and apparently was never legally named Newark Watch Company.[3] Arthur Wadsworth was an English watchmaker whose son later started the Wadsworth Watch Case Co. in Dayton, Kentucky. He had been head watchmaker for Fellows & Schell, designed the Newark movement, and patented their stemwind mechanism. As movements reached the market in 1867, they were among the earliest American stemwinders. Some were marked Keyless Watch Co., New York, as several lever-set stemwind mechanisms were tried without success. The mechanisms were either faulty, or more likely, uneconomical since additional parts and special cases were required. Most Newarks were therefore keywind, and many were marked with jewelers' names. Quality was never notable, so that the watch probably never established a good reputation. In 1869 watches were still not being sold at a profit, so Robert Schell became pressed to sell the factory. This opportunity

41. Elgin Ryerson #7081 — inexpensive 7-jewel grade ca. 1867.

42. Early view of machine shop at Elgin ca. 1870.

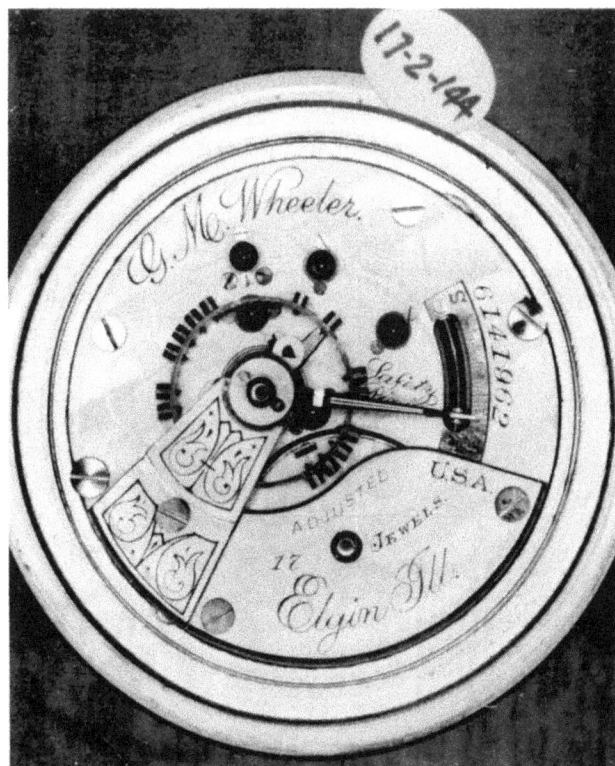

43. Elgin Wheeler #6,141,962 — 17-jewel railroad grade ca. 1895.

44. Elgin Father Time #15,807,079 — 21 jewel split plate railroad grade ca. 1910.

45. N. B. Sherwood — instrumental in establishing the shop of the Newark Watch Co.

46. Hibbard #20,946 — Cornell Watch Co. ca. 1872.

prompted J. C. Adams to form another watch company, this time with Paul Cornell, who had real estate dealings in the Chicago area. They set up the Cornell Watch Co. and in 1870 moved what was basically the Newark Watch Co. into a new factory at Grand Crossing, Illinois, near Hyde Park, to produce what was basically the Newark watch. The identical watch brought identical results so that by 1874 Mr. Cornell was also forced to sell the operation. Hearing of cheap Chinese labor in California, he convinced William Ralston, cashier of the Bank of California, to enter into a partnership and move the factory to San Francisco. Thus began the Cornell Watch Company of San Francisco, with no Chinese labor because of the reaction from the white labor force, still producing the old Newark watch, which continued to prove unpopular. Thus ended the company in 1876, followed by the almost instantaneous episode of the California Watch Co. in that same year, along the same story line. Some California watches were finished at Chicago in 1877 under the name of the Western Watch Co. and with that the Newark watch came to the end of a brief but busy career.

No further description is necessary to point out the fate awaiting those who ventured into watch manufacturing unprepared for the effort and expenditures required to make such a factory work. Beside capital to launch into business, a company needed the reserves to introduce new models and keep up with current fashions, for the competition was fierce. Everybody's chief competitor was the American Watch Company, which by 1870 was large, efficiently mechanized, and boasted a progressive history of introducing nickel finish, stemwinding, and thinner 18 size movements, as well as 20, 16, 14, and 10 sizes to suit the tastes of both ladies and gentlemen. The National Watch Co. was close behind and growing rapidly, while E. Howard & Co. was building a strong reputation as a strictly high-grade watch-

maker. By contrast, the Newark company and its successors made almost solely an 18 size keywind watch that was going out of fashion before the first one left the factory.

Newark's twin, the United States Watch Company, was the offspring of Giles, Wales and Company, a New York jewelry house that imported English and Swiss watches and had enjoyed considerable prosperity during the Civil War.[4] Tempted to try watchmaking they hired James H. Gerry from the American Watch Co. in 1863 to start a machine shop. Gerry had been a member of the Nashua watch venture and, as a watch mechanic in demand, probably joined more watch companies than any other individual in the industry. He brought to United States a band of machinists from Waltham and got the shop going in Newark, New Jersey. Giles, Wales & Co. decided to form a stock company in 1864, building a new glass and iron factory in Marion, New Jersey, that was occupied the following summer. An 18 size fullplate movement was designed in the English style by a New York watchmaker named Oliver Baldwin, with the "butterfly" opening in the top

47. Factory of the Cornell Watch Co. ca. 1872.

MANUFACTORY OF THE UNITED STATES WATCH COMPANY.

49. Factory of the United States Watch Co. of Marion, New Jersey, ca. 1870.

48. Frederic Atherton & Co. #1360 — United States Watch Co., Marion, NJ, ca. 1867. (Photo courtesy of NAWCC Museum, Inc.)

50. Fayette Stratton #8241 — United States Watch Co. ca. 1870.

51. Tremont Watch Co. #9095 ca. 1866.

plate patented by Mr. Giles to check the escapement action in the assembled watch. By 1867 the first of these left the factory and about the same time James Gerry had a disagreement with Mr. Giles and resigned, much like N. B. Sherwood at Newark. Much to their credit, the United States Watch Co. made some very fine watches that were among the earliest in America with stemwinding, nickel finish, and damaskeening. They also produced an extraordinarily fine ¾-plate movement. Unfortunately, the cost of all this work, plus lavish spending on the factory and grounds, never permitted sales revenue to approach cash outlay. The company reorganized as the Marion Watch Co. in 1872 in an attempt to recover, but this failed in 1874, after which a final attempt as the Empire City Watch Co. closed permanently in 1877.

The Tremont Watch Co. was a clever attempt by the industry founder, Aaron Dennison, to avoid the expense of starting a new company.[5] Seeing good market prospects, he organized the company in 1864 with A. O. Bigelow of Bigelow, Kennard & Co. jewelers of Boston. To circumvent the high price of skilled labor in America, he concluded, it would be more economical to make trains, escapements, and balances in Switzerland, while making the plates, barrels, and flat steel parts in Boston. Belding D. Bingham, organizer of the Nashua Watch Company, was put in charge of the Boston manufacturing operation while Dennison went to Switzerland. From a shop in Zurich, Dennison demonstrated to the skeptical Swiss that he could produce 600 sets of finished material per month, accumulating parts faster than the group in Boston could use them. In 1865 the first complete watches were ready for sale, again being standard 18 size fullplate, gilt movements, and the whole business was looking like a successful plan. As opposed to those who wouldn't update an obsolete idea, the backers of the Tremont Watch Co. couldn't resist changing something that worked. Inspired by modest success, they were encouraged to increase their grasp of the business, deciding to manufacture their entire watch in the Boston suburb of Melrose, where two of them owned property. This was much to the dissatisfaction of Mr. Dennison, who was all too aware of the expense of manufacturing the en-

tire watch in America, so he departed the company and completed the last Tremont parts by contract. Watches began leaving the Melrose Watch Co. in 1867 or 1868 and the group began work on a ¾-plate model. Unfortunately, the cost of this consumed all available funds and the stockholders simply closed their operation in 1868 by mutual agreement. Aaron Dennison was contacted to find a buyer for the machinery, which he did in Birmingham, England, where it was set up as the English Watch Company. Dennison was not part of the English company, which struggled briefly to a close, but remained near Birmingham where he started a watchcase company which operated into the 1960's.

52. Tremont Watch Co. #32,299 — Melrose Watch Co. model ca. 1867.

53. Tremont Watch Co. ¾ plate #4346 ca. 1867.

The Dennison Watch Case Manufacturing Co. was started by Aaron Dennison in Handsworth, England, at age 62. Although there were borrowings and assistance in organizing the company, it was owned and managed by Dennison alone. He applied the same type of machine manufacturing techniques to watchcases as he used for watchmaking. The company was successful and became well known in the world jewelry trade. Dennison's creativity had previously led to the Dennison Mfg. Co. (paper products) and the American Watch Company, both of which became prosperous, and neither of which netted Dennison any financial gain. With his last venture he finally reaped the profits of his own inventive skill. The founder of the American watch industry remained in England till his death in 1894 at the age of 83, and was buried in the local church yard at Handsworth near the grave of James Watt.[6]

As a sidelight, it is interesting to speculate that Dennison's Tremont Watch Co. may have provided a model for an all Swiss "American" watch, produced on the American system with cheaper European labor. Whatever his inspiration, Florentine A. Jones left his position at E. Howard & Co. in 1868 and rented space in Henri Moser's new water-powered factory in Schaffhausen, Switzerland, to start the International Watch Company. He wanted to make a good quality watch but thought it could be done more economically than at Howard where a number of hand-finishing techniques were used. Taking Dennison's approach a step further, he reportedly made the entire watch on American machinery but with Swiss labor.[7] He was forced to sell out in 1871 but the business continued under new ownership and is still operating. Perhaps coincidentally, the Philadelphia Watch Co. was also formed in 1868, incorporated in Philadelphia by Eugene Paulus, who had worked there as a watchmaker for a number of years. He had patented an escapement in 1859 plus the plate design and winding click for the Philadelphia watch in 1868. The company marketed a watch similar in appearance to the Howard series III and advertised through the 1870's as making their watches with intelligent labor on American machinery.[8] While they maintained a business office in Philadelphia, no factory site has ever been identified. The circumstances suggest the Inter-national Watch Co. and a scheme by which the young company might have sold their Swiss watch in the United States without disclosing its source. In any event, the Philadelphia Watch Co. marketed ¾-plate watches in a full range of quality. Some finishing and adjusting may have been done in America but movements were of Swiss manufacture with a distinctly American appearance.[9]

Dietrich Gruen (1847-1911) may also have been influenced by Dennison's Tremont Watch Company, or by the later Philadelphia and International companies. Having come to America from Germany, he began a watchmaking plant in 1874, a period when the Philadelphia Watch Co. was advertising widely. Swiss ebauches made to order were finished in his Columbus, Ohio, shop, incorporating his patented safety pinion. By 1879, business warranted expansion into a partnership known as Gruen and Savage, the latter being W. J. Savage who handled the commercial aspects while Gruen oversaw manufacturing. This organization formed the nucleus of the Columbus Watch Co. in 1882, which fabricated completely American-made watches. In 1894 Gruen left Columbus, combining with his son, Fred, to resume a business selling imported movements as D. Gruen & Son. Eventually, Gruen purchased manufacturing interests in Switzerland that supplied material to finishing facilities in Cincinnati, Ohio.[10] This proved to be a successful business that was one of the longest surviving American watch companies. It was likely an inspiration for such international ventures as Benrus and Bulova, which were a later reflection of Dennison's genius. Beside American watchmaking, Dennison originated multi-national industry, which he lived to see Gruen make fashionably profitable.

The final company of 1864 was the brainchild of Don Mozart, an Italian immigrant who was something of a mechanical genius, if a bit misdirected. After various wanderings and mechanical pursuits, he had gone to Bristol, Connecticut, in 1853 to manufacture a clock going for one year, which proved a failure. By 1864 he had worked out details for his watch, which included his unique escapement and a three-wheeled train, already used in English watches at the turn of the century but probably original with him. He found financial backers in the prosperous New York jewelry trade and in 1864 the New York Watch Co. set up its

54. Philadelphia Watch Co. #1867 — American company with an American-looking movement made in Switzerland.

55. Early Columbus Watch Co. movement made from Swiss material of Gruen and Savage Company. Contains Gruen's patent pinion and bears Gruen's serial number.

56. Factory of New York Watch Co. in Springfield, Mass. Burned in 1870.

58. Springfield, Mass., factory of Hampden Watch Co. ca. 1880.

machine shop in Providence, Rhode Island, with Don Mozart as superintendent.[11] Mozart had underestimated the job and by 1866 the backers were becoming exasperated, for watches still seemed a long way off. At that point Mr. L. W. Cushing was brought in from Waltham to reorganize the machinery for a conventional American watch, and Don Mozart left the company for Ann Arbor, Michigan, where he began anew. Only a short time later the Providence company had the opportunity to acquire a vacant factory in Springfield, Massachusetts, on convenient terms, and moved there in 1867. As they occupied the factory, they hired James H. Gerry from the United States Watch Company, who brought with him some of his itinerant Waltham machinists. A fire destroyed the building just as watches were becoming ready for sale in 1870, but much of the machinery was saved. Recovering rapidly, watches did reach the market in 1871 and the little company finally got on its way. They were one of the earliest companies, beside Waltham, to concentrate on the ¾-plate design, producing some excellent timepieces over the next few years. Business degenerated when financial panic gripped the country in 1873. The stockholders reorganized as the New York Mfg. Co. in 1875 but that company managed to survive for only another year. A second reorganization by the same stockholders started again as the Hampden Watch Company, named after the county where Springfield is located. Incredibly,

the company began to thrive, for the recession was easing. In 1888 they merged with the Dueber Watchcase Co. of Newport, Kentucky, and the combined firm moved to a large new factory in Canton, Ohio. Production was never large compared to Waltham or Elgin but they made a number of commendable models, including the first American 23-jewel watch. They continued till 1930, when the entire factory was purchased by Russia to establish watch manufacturing in that country. Like many before, Russia found it easier to buy watchmaking technology than invent it all over again.

This burst of new companies in 1864, some of which began operations in 1863, initiated the expansion phase of American industrial watchmaking. Most individuals instrumental in forming these organizations were businessmen wishing to cash-in on proven technology, not craftsmen striving to advance the Industrial Revolution. Their entrepreneurial spirit was a clear indication that pioneering was ending and profiteering had begun. Their zeal was inspired by financial success of the American Watch Co. in Waltham, which prospered under Royal E. Robbins' direction. Dennison and Howard had pioneered technology for factory watch manufacture; Royal Robbins built watch manufacture into a lucrative business. Robbins' success fortified a band of followers and promoted industry growth by introducing innovations that broadened the watch market. The strength and character of his company were dominating factors for this new industry, making Robbins one of the most important individuals in shaping American watchmaking.

Conditions were right for strong growth in watchmaking, even though the overall economy was weak. Watches had previously been luxury items, so that most people desired one. Rapidly falling prices brought watches within reach for more people, and with every drop in price a larger market opened to watch producers. Booming immigration further boosted the sales market, in addition to which all kinds of industries were growing, so that an increasing portion of the population earned regular wages rather than being in agriculture. This all left room for a number of watch companies, while cash flow from increased sales volume financed further expansion. The most obvious expansion was this growth of sales and proliferation of companies, and watch tool companies were appearing nearly as rapidly as watchmakers. In addition, sophistication and diversity of watches themselves were expanding into every conceivable area:

a) precision timekeepers;
b) complicated watches such as chronographs;
c) inexpensive jeweled watches;
d) cheap unjeweled watches;
e) stemwinding and setting;
f) nickel damaskeening;

57. King #5641 — early high-grade ¾ plate movement of New York Watch Co. ca. 1870.

59. Rice #23,277 — New York Watch Co. ca. 1874.

60. Canton, Ohio, factory of the Hampden Watch Co. ca. 1890.

g) smaller watches for both ladies and gentlemen;
h) thinner watches.

Expansion into precision timekeeping was essential for the growing watch industry in order to improve machine capability and to establish public credibility. Once it had developed and demonstrated technical capability to produce precision watches, the industry enjoyed an indisputable reputation for quality. This reputation attracted millions of customers as cheaper and more diverse watch grades appeared at popular prices.

Industry expansion into these new watchmaking fields was based on corresponding increases in watch tool capability. Machines and techniques had to be developed for manufacturing higher precision and new kinds of parts, as well as machines that would produce the old types of watches faster and more efficiently. It was no easy task to make thousands of quality parts each day. Growth of machine technology was really the essence of expansion in watchmaking, for machines were the muscle of factory production. Foreign competitors continued hoping, through the 1870's, that machines were too clumsy to make high-grade timekeepers, or too expensive to underprice cheap hand-finished watches. Both dreams had been shattered when the period of expansion closed in 1880. Handcraftors were then suffocating as American machinery relentlessly fabricated all grades of watches by the thousands; thereafter the machines had only each other as competition.

The extremes of machine capability were demonstrated as American factories exhibited mass-production of quality watch parts at the Philadelphia Centennial Exposition in 1876, and mass-production of dollar watches in 1878. Americans made an impressive showing at the Centennial by operating automatic machines on the premesis and exhibiting premium-grade watches that demonstrated quality surpassed by few foreign timekeepers. Shortly thereafter, dollar watches undermined inexpensive hand-finished watches that supported the Swiss industry. The impact of these events was to expand machine watchmaking into another important area: Switzerland.

Through the 19th Century, Swiss mechanics had gradually developed machines for performing certain watchmaking tasks, while their industry remained primarily a handcraft. By 1880 it was clear to them that a completely machine oriented system was required to remain in competition with America, and they thus began to reorganize their industry. Diverted by this retrenching effort, a diminished Swiss presence was felt in America during the last quarter of the 19th Century, except in the limited market for complicated high-grade watches, where handwork was more justified than the massive tooling expenses required for machine production. The new Swiss industry appeared around 1900 sporting sleek little wristwatches, which were not readily accepted in a rugged country typified by the wild west and Teddy Roosevelt. Another ten years elapsed before wristwatches became a significant influence. The tables were then turned and the American watch industry faced a sophisticated competitor using a well developed system of machine manufacture. So, the American industry, through its Age of Expansion, spawned a foreign industry as well, and may thereby have hastened its own demise.

Table III
Total Production Quantities 1880

Conventional Jeweled Watches

Waltham	1,500,000
Elgin	700,000
Hampden	200,000
Illinois	190,000
Lancaster	50,000
Rockford	50,000
Howard	50,000
U.S. Marion	20,000
Newark	13,000

Dollar Watches

Waterbury	100,000

REFERENCES

1. Charles S. Crossman, *The Complete History of Watchmaking In America* (reprinted Adams Brown, Co., Exeter, NH), p. 91.

2. Ibid., p. 70.

3. Dr. Percy L. Small, *Discussion of Dr. W. Barclay Stephens' Paper "The Newark Watch Co. and Its Career,"* (BULLETIN, National Association of Watch and Clock Collectors, December 1950), p. 256.

4. Crossman, p. 83.

5. Ibid., p. 104.

6. Henry G. Abbott, *A Pioneer* (reprinted Adams Brown Co., Exeter, NH), p. 38.

7. T. P. Camerer Cuss, *The Country Life Book of Watches* (Middlesex, England, The Hamlun Publishing Group Ltd., 1967), p. 101.

8. William Muir, *The Problem of Chestnut Street* (BULLETIN, National Association of Watch and Clock Collectors, Inc., August 1969), p. 1019.

9. Larry Treiman, *The Philadelphia Watch Company — Revisited* (BULLETIN, National Association of Watch and Clock Collectors, Inc., December 1978), p. 597.

10. Eugene T. Fuller, *The Priceless Possession of a Few* (BULLETIN, National Association of Watch and Clock Collectors, Inc., Winter Supplement 1974), p. 8.

11. Crossman, p. 108.

SECTION 2
OPERATION OF THE INDUSTRY
IV
THE AVERAGE AMERICAN WATCH

After the companies of 1864, there was a hesitation in American watchmaking as businessmen waited to see how this group would fare, and the only completely new company formed in the next decade was Illinois Springfield Watch Company. When additional firms began appearing in 1874, the nature of the industry had been solidifying for 15 years: popular watch styles were set, fundamentals of watch machinery were established, and a group of machinery designers had become known. Frequently, new companies hired recognized mechanics from other factories, purchased used machinery, and fell in line with popular trends of watch design, which all continued to promote a uniform style of American watchmaking. By 1880, when the age of competition began, factory watchmaking was hardly an experimental industry. There was a fully developed competition to produce the Average American watch in large quantity, in decent quality, and in a profitable manner. Various new companies pursued this goal in a distinctly unvaried manner, in an effort to lower prices and gain public acceptance. Watchmaking technology had become common by 1880, so that chief tools of the competition were managerial and marketing skills. The expense of developing new products had to be carefully considered, for tight budgets began trimming the capital available for long term investment as a result of thinning profit margins. Driven by Elgin and Waltham, prices across the watch trade continued to fall. It naturally followed that changes in watchmaking slowly occurred by evolution, and were frequently resisted by existing companies. However, three events prompted rapid change within the American watch industry:

a) introduction of inexpensive "dollar watches" in the late 1870's;
b) establishment of guidelines in 1892 for railroad watches;
c) rising popularity of wristwatches after World War I.

Cheap watches brought lower prices and stimulated invention of sophisticated automatic machines capable of high-rate production. Railroad watch standards brought forth new high-grade movement styles on one hand, but on the other, these types of railroad watches instantly became uniform in style themselves. The small size of wristwatches created a technological crisis within American watchmaking, requiring a large investment to educate and retool for the more stringent requirements of these tiny mechanisms.

Because of the tendency to hire known mechanics and machine designers, one can trace numerous careers through the watch factories of America, from the very beginning. Nelson Stratton, who had been a Pitkin apprentice in 1838, joined Dennison's Boston Watch Company, started the Nashua Watch Company in 1859, and worked for the Waltham company, where he was London agent for a number of years. James Gerry became the classic itinerant watchmaker, for by 1870 he had been at the Boston, Nashua, American, United States, and New York watch companies. He continued on to work for Howard, Auburndale, a clock business in Elgin, Illinois, and the National Clock Co. of Brooklyn. Of ten individuals who left Waltham to start the National Watch Co of Elgin, half moved on to the Illinois Springfield Watch Co. after 1869. This trend continued throughout the years of American watchmaking. The dis-

61. Elgin National Watch Co. factory ca. 1890.

62. Howard #51,421 — key-wind ¾ plate ca. 1865.

slowly modified. Most new trends were introduced during the 1870's although some did not make a general appearance till a number of years later. Table IV lists introduction dates for several features, showing that the 1870's were a busy decade. Since the industry was still striving to establish its reputation during this period, some companies built a limited quantity of carefully made watches incorporating all of these improvements. Wishing to demonstrate that factories had become capable of manufacturing high quality, they took time to build exquisite timepieces at exquisite prices beginning around $100. The competitive rush that mounted during the 1880's reduced the level of luxury that companies could afford to expend. Even so, as production quantities increased, the industry was still able to incorporate techniques it had developed for producing fine detail and high quality. Improvements spread to common grade movements yet prices held steady and even dropped. When railroad watch standards popularized higher-grade models, the industry was easily able to manufacture enough for everyone, not just the rail trade. As a result of all these influences a 17-jewel stemwind could be purchased in 1910 for under $10. Dennison's original Boston watch had been a fullplate keywind movement, 18 size with gilded finish, 7 to 15 jewels, lever escapement, and most often a hunting case. Of these traits only the lever escapement endured, so that the common watch of 1930 was far from that of 1850.

Keywinding was the first of the old traits to disappear. English watches had nearly all been keywind because fusee mechanisms, which English watches usually had, made stemwinding very difficult. Several stemwind devices for going barrels, which American factories mostly used, had been invented in Switzerland during the 1840's but when the American industry was beginning in 1850 these were still not widely utilized. This was just as well since the

astrous Cornell Watch Co. employed George Clark, who became superintendent of Aurora, J. W. Hurd, superintendent at Aurora after Clark, Albert Troller, later superintendent of Rockford, and John Logan who went on to form his own spring company. When Seth Thomas started making watches, they brought in H. Reinecke of Waterbury Watch Co. and C. Higgenbottom of Hampden. The Cheshire company hired D. A. Buck, chief inventive brain of Waterbury Watch Company. The Trenton Watch Co. employed men from many New England watch companies as heads of departments. Auburndale hired Chauncy Hartwell of Waltham while also having on the staff William Wales, a founder of United States, James Gerry, and E. H. Perry, co-founder of Adams & Perry.[1] Profit was more important than originality when forming a company, so the prime motive in staffing the shop was to hire people who knew mechanical watchmaking. The capital required to start a manufacturing business was tremendous since this staff would have to be supported for some time before any watches would be sold. Likewise, machine tools were immensely expensive and second-hand machinery could usually find a buyer. Many machines traveled as often and far as their designers before ending in a scrap heap or factory fire. These costs were among the chief deterrents to inventive sampling in watchmaking. New ideas had to be based on something the wholesale trade, retail jewelers, and customers would all buy, and success depended on establishing a sound reputation around a desirable product and selling it in quantity.

Prior to the American industry, English watches were preferred, so Dennison shrewdly built American movements with an English appearance. His standard sizes offered customers many combinations of cases and movements, and interchangeable parts offered jewelers a ready supply of replacement hardware. For 30 years the average watch remained as Dennison conceived it and even after that it faded only gradually, as its characteristics were

63. Waltham model 70 #501,090 ca. 1870. Early stem wind with setting button edge of case, the latter feature being unusual on American watches.

38

infant industry had enough problems without the added complication of stemwinding. Considerably more capability was available in the 1870's so that stemwinding began appearing in finer-grade movements and was universal by the mid 1880's. Keywind mechanisms had all been similar since little could be done to something so simple. The complexity of stemwinding offered ample room for variation and the industry created numerous designs covered by a maze of patents, though most of these were derived from the two basic types developed in Switzerland. The rocking-bar style was invented by Antoine LeCoultre during the middle 1840's and was most often associated with lever set watches.[2] It appeared in many 18-size movements and in 16-size movements for the rail trade. For non-railroad grades it was most common to find a shifting-sleeve mechanism, invented by Adrien Phillippe in 1842.[3] With this design the hands were set by pulling out the stem, which was more convenient in everyday watches. Since both types involved many steel parts, considerable effort was invested in developing simple yet dependable designs.

Table IV

PROGRESSION OF THE AMERICAN WATCH

1850   First factory — Dennison Howard & Davis
1853   First Boston Watch Co. pieces — fullplate, 18 size, gilded, keywind
1858   Chronodrometer — Waltham
        18,000-beat train — Howard
1860   Split-plate movements — Howard
1861   ¾ plate, 18 size — Waltham
1863   20, 16, 10 sizes — Waltham
1864   Factory production spreads
1865   Safety pinion — Waltham
1866   First specifically-made railroad watch — Waltham
1867   Damaskeening — U.S. Marion
        Stemwind, lever setting — Newark
        Stemwind, button setting — U.S. Marion
        First Elgin — B. W. Raymond, railroad model
1868   Stemwind, lever setting — Waltham
1869   Railroad model "70" — Waltham
1870   14 size — Waltham
        Reed's whiplash regulator — Howard
        Stemwind, pendant setting — Howard
1873   6 & 8 sizes — Waltham
        Stemwind — Elgin
1874   17 size, cased — Elgin
1875   12 size — Waltham
        14 size — Elgin
1876   Cheap watches — Auburndale
1877   Chronographs — Waltham
        Nickel plates — Elgin
1878   16 size convertibles — Elgin
        Longwind cheap watch — Benedict & Burnham
1883   Inexpensive jeweled watches — Cheshire
        Breguet spring production — John Logan, Waltham
1885   Pendant setting — Waltham
1892   Railroad watch standards
1896 $1 watch — Ingersoll
1908   Last factory — Manistee, Michigan

The earliest American stemwinders were both button and lever set. Newark and U.S. Marion introduced stemwinding as their first watches began selling in 1867. Waltham responded with a stemwind design in 1868. Each required a lever on the edge of its case to be pulled out in order to set the hands. All were rocking bar type mechanisms, though Newark used one design in which the bar translated rather than rocked. A few American watches, by U.S. Marion and Waltham, utilized button setting, in

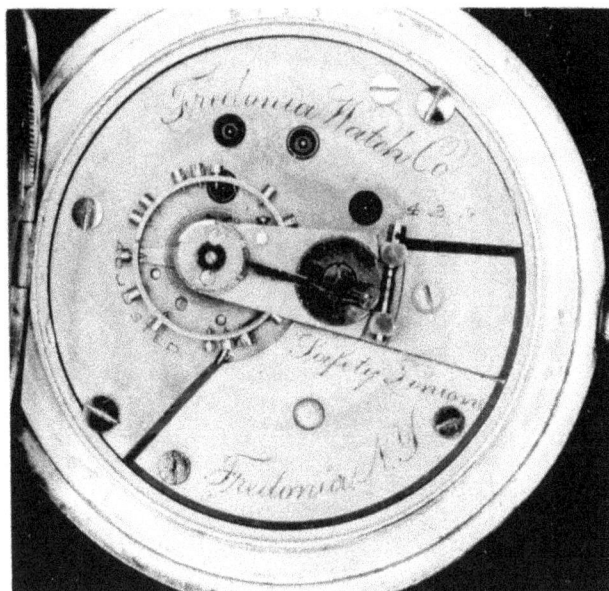

64. Fredonia #429 ca. 1883 — has unusual worm gear regulator.

which a button on the case was pushed when hand setting was desired. Though button, or push, setting was common in European watches, American companies generally preferred a pull-out lever. New York Watch Co. (Hampden) and Adams & Perry both offered stemwinding in their early models, and by 1880 stemwinding dominated new watch sales. Thanks to machine production, winding mechanisms had been added to common watches without significantly increasing prices. Edward Howard introduced stemwinding in 1869 or 1870, using a shifting-sleeve mechanism invented by James Gerry that proved to be highly reliable. This was America's first stemset watch, in which the stem was merely pulled out to set the hands. Sure-acting stemset devices were difficult to achieve, which may account for why Waltham did not market stemsetting till its 88 model

65. Aurora #80,639 ca. 1887. 15-jewel railroad grade. Note fifth pinion pivoted near regulator scale.

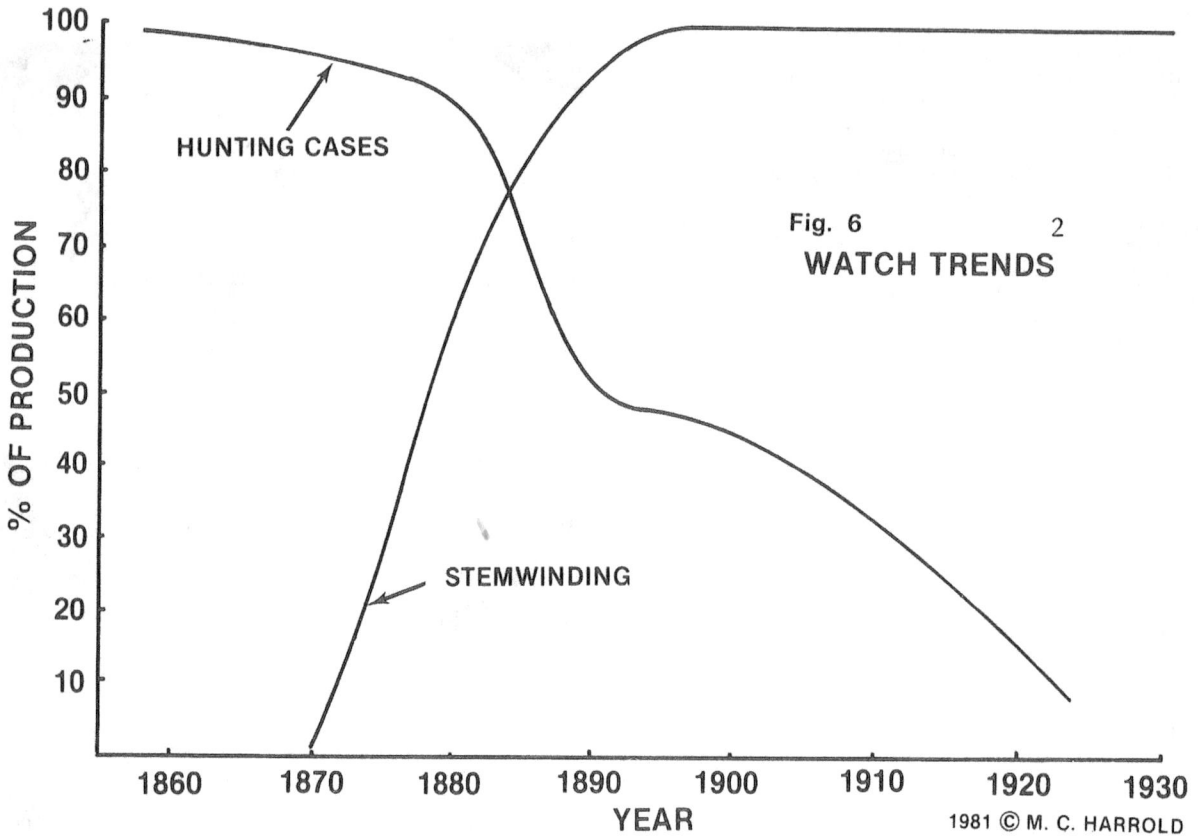

Fig. 6
WATCH TRENDS

HUNTING CASES

STEMWINDING

% OF PRODUCTION

YEAR

1981 © M. C. HARROLD

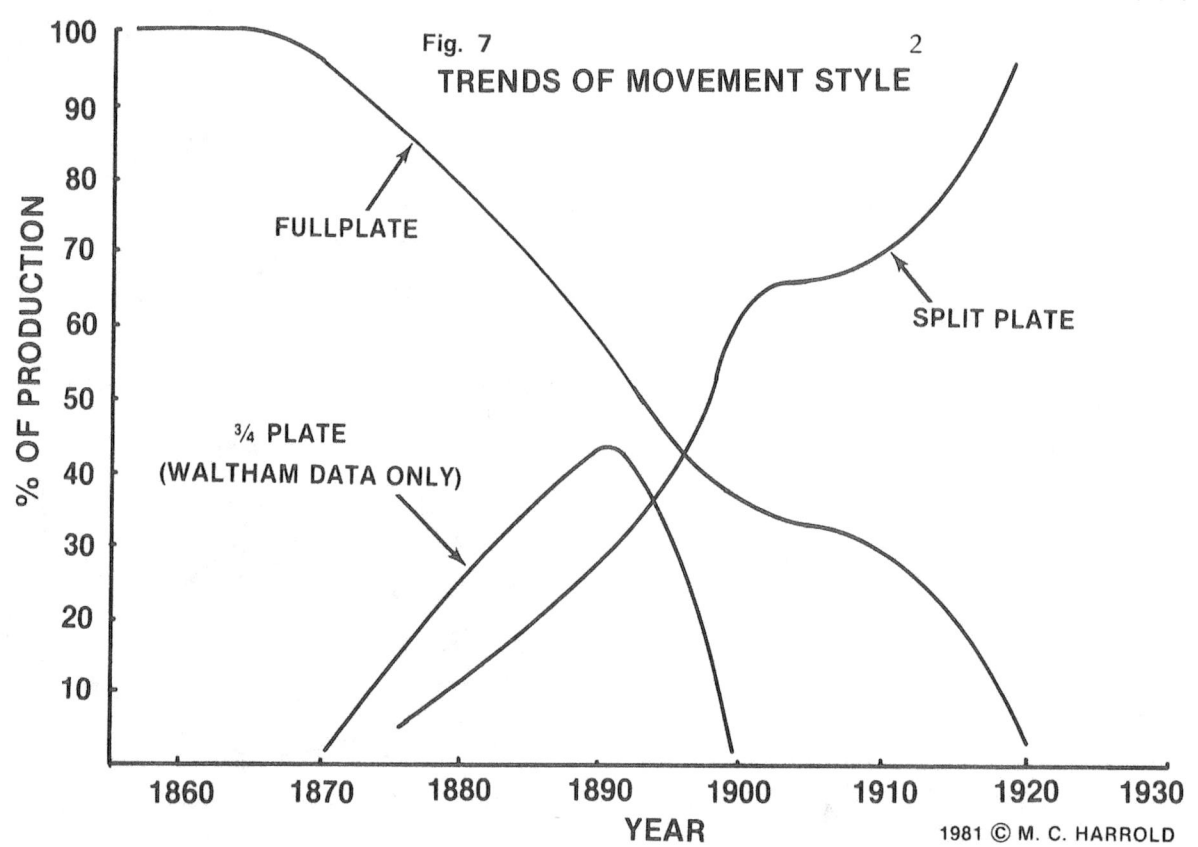

Fig. 7
TRENDS OF MOVEMENT STYLE

FULLPLATE

SPLIT PLATE

¾ PLATE
(WALTHAM DATA ONLY)

% OF PRODUCTION

YEAR

1981 © M. C. HARROLD

66. Seth Thomas #229,174 ca. 1893.

68. Waltham #6,506,461 ca. 1895 split plate model 88, similar in appearance to ¾ plate.

67. Elgin #6,144,577 ca. 1895.

69. Rockford #381,749. 17-jewel split plate ca. 1898.

70. Columbus #364,165. 17-jewel railroad model ca. 1895.

72. Waltham #15,105,609 split plate ca. 1909 — separate-appearing bridges are formed from one piece.

71. Rockford #593,982 split plate ca. 1905 — separate-appearing bridges are formed from one piece.

73. South Bend #1,198,104 split plate ca. 1925 — separate-appearing bridges are formed from one piece.

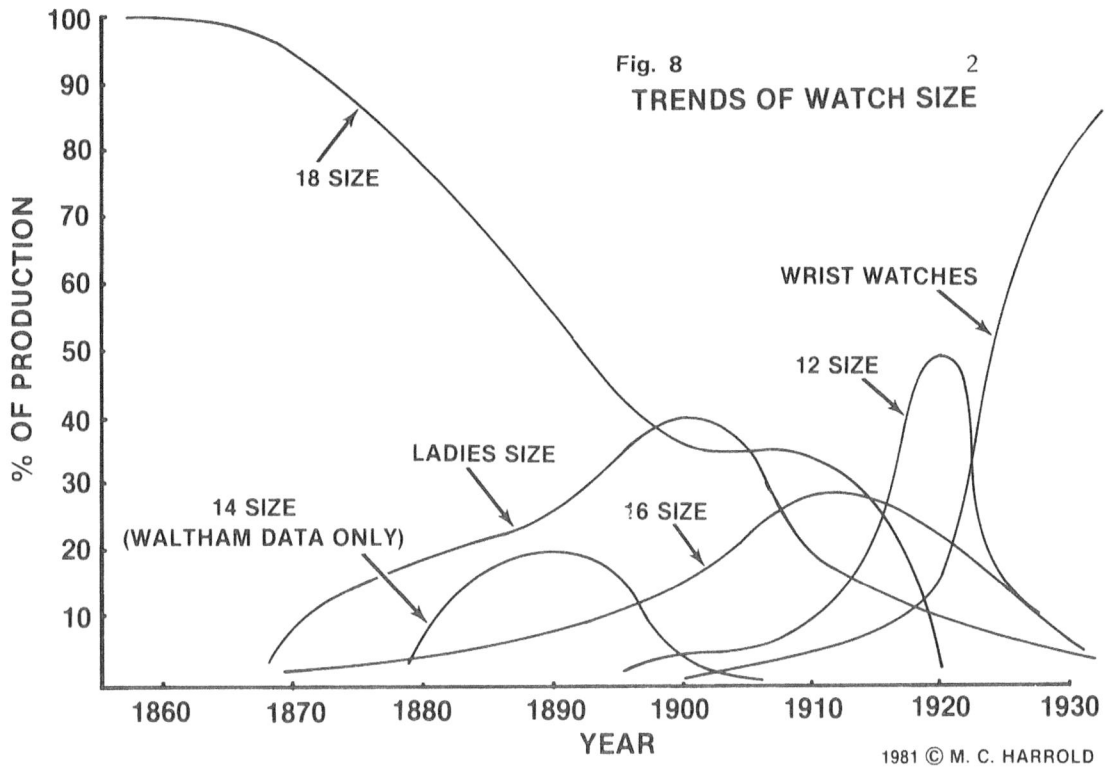

Fig. 8   2
TRENDS OF WATCH SIZE

% OF PRODUCTION

18 SIZE

WRIST WATCHES

12 SIZE

LADIES SIZE

14 SIZE
(WALTHAM DATA ONLY)

16 SIZE

YEAR

1981 © M. C. HARROLD

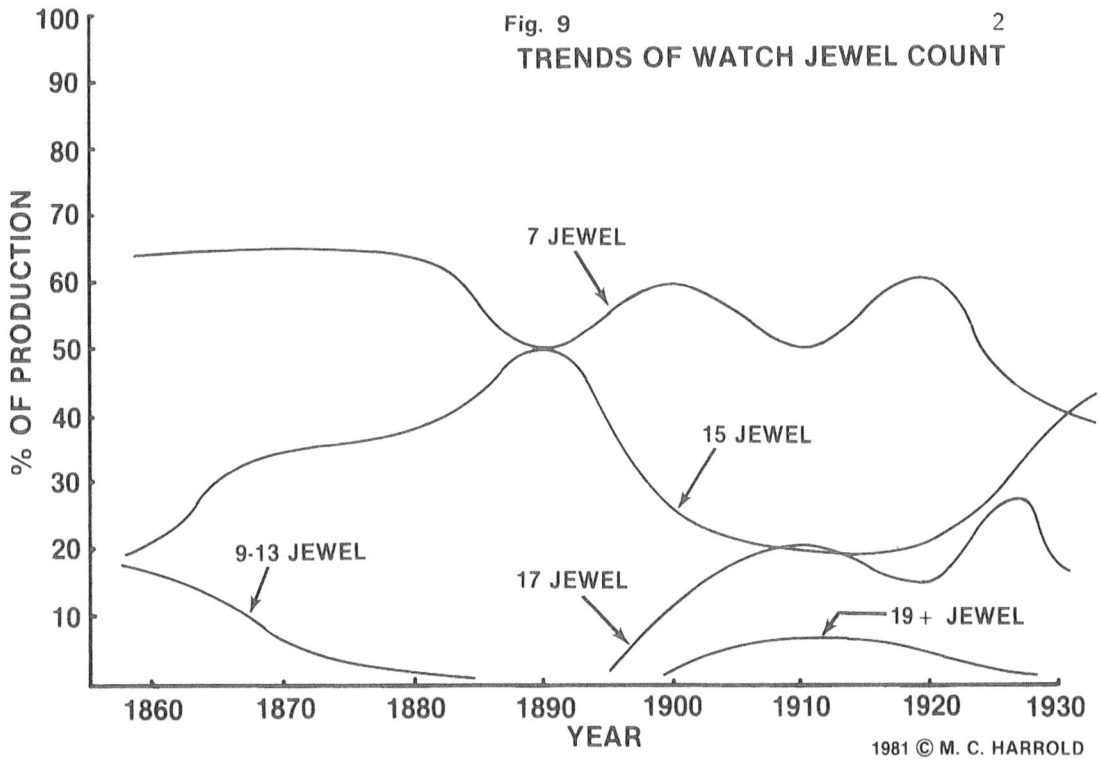

Fig. 9   2
TRENDS OF WATCH JEWEL COUNT

% OF PRODUCTION

7 JEWEL

15 JEWEL

9-13 JEWEL

17 JEWEL

19 + JEWEL

YEAR

1981 © M. C. HARROLD

(1888). The Waltham 88 was not a wholly satisfactory stemset design, and was replaced by an improved device in later Waltham models.

The fullplate style lasted for 60 years as the "man's watch" since it presented an appearance of massive ruggedness. It was the least expensive design to manufacture and after 1900 represented the lowest price models from conventional jeweled watchmakers. It was also thick, so ¾-plate and split-plate movements were introduced, becoming thinner by placing the balance down beside the train instead of on top of it. Nashua Watch Co. was first with ¾-plates in America, copying them from the English, and Waltham developed them to some prominence. Adams & Perry, New York Watch Co. and U.S. Marion were the other companies to adopt ¾-plates, in each case using them for high-grade movements in low quantity. Waltham committed heavily to ¾-plate designs, many with the plate trimmed back to minimum size, but fullplate movements continued to dominate men's watch sales till railroad watch standards brought forth split-plate styles after 1890. At that point Waltham yielded to public taste by dropping ¾-plates completely: Split-plate movements were the final approach to pocket watches, supporting the barrel and train under two separate bridges. True bridge models, with individual bridges for each wheel, were seldom made in America and models with this appearance were usually split plates with "bridges" formed from one plate. This was because separate bridges were more expensive. A fullplate could simply be turned on a lathe, split plates could be turned and contoured, but bridges meant contouring numerous odd shapes. Though slightly more expensive than fullplates, split-plate movements were easier for repairmen to work with and therefore more popular in the retail jewelry trade. By 1900 they were firmly in command of the watch market and by 1920 were virtually the only style made.

Large 18 size watches were synonymous with fullplate styles, lingering till 1920. The early industry tried 20 sizes but these were inconveniently large and the general trend was in the other direction. There was a shift down to 16 sizes when thinner ¾-plate models were introduced and 16 size watches were never made in a fullplate style. Split-plate 16 size movements increased sharply in popularity after the railroad standards in 1892, enjoying a good share of the market for another 40 years. Elgin tried marketing some inexpensive 17 size movements during the 1870's but these were soon overshadowed by dollar watches. They were some of the few conventional jeweled watches to sell in factory cases till Keystone brought out cased watches under the Howard name after 1900. At that time 12 sizes were becoming popular, representing the trend for gentlemen to carry progressively smaller more streamlined watches. The 12 size then dominated men's watches till wristwatches took over in the late 1920's. Waltham made an effort to promote 14 size watches in the 1880's but many went to England and they never sold heavily in America. To attract the ladies' market, early companies introduced 10 size movements during the 1860's which remained popular for some years. By 1890 this market was taken over by 6 sizes and after 1900 the 0 size was popular. Ladies' watches thus followed men's styles toward smaller size.

The earliest watches of Dennison and Howard had used large clear aquamarine jewels in their top plates through which the pinions could be seen rotating below. These were known as Liverpool windows after watches made in that English city during the 1820's and 30's, but were soon dropped in favor of small plain garnet jewels. These were usually burnished into the plate on lower-grade movements; i.e., a shoulder was machined into the watchplate to receive the jewel and a small amount of metal was turned down over the jewel to retain it. This was a perfectly effective method long used by Swiss makers but presented difficulties if the jewel needed repair. Extracting a broken jewel from a plate and setting a new one was a job few workmen could accomplish without leaving an ugly scar. Better movements followed the English practice of placing each jewel in a metal setting which in turn was fitted snugly into the plate and held with screws. These could be removed and replaced with ease. Beside aiding repairmen, jewel settings appealed to Americans, who liked having lots of jewels in their watches. English and Swiss makers usually used no more than 11 jewels, except in watches intended for the American market. Indeed, even half the American-made watches were plain jeweled, which is to say a basic 7-jewel watch. The remainder tended to drift upward in jewel count. In early years 9, 11, and 15 jewels were common, progressively becoming mostly 15 jewels. Railroad standards required 17 jewels minimum, so that 17-jewel models began replacing 15-jewel movements during the 1890's. Watches of 19 jewels or more also appeared but tapered off after 1910 as the railroad market began saturating. Most highly-jeweled watches had jewels in settings and finer grades used gold settings standing above the plates.

The basic 7-jewel watch can be seen in Figure 10. An 11-jewel watch had the basic 7 plus hole jewels on the lever and escape wheel arbors. As the number increased from 13 to 15 to 17, hole jewels were added to the 4th, 3rd and center arbors respectively. Watches were occasionally made with jewels on the top plate only, giving the appearance of being more highly jeweled than they really were. From this one can sense the importance people attached to seeing jewels when they opened the backs of their watches. Through 1890 many smaller watch companies imported finished jewels from Switzerland but large firms imported raw jewels and finished them in their factories. Jewels were generally ruby, sapphire, garnet or aquamarine, though diamond was occasionally used for cap stones which had no holes through them. Some garnet was obtained from Montana but mostly imported from Bohemia while ruby and sapphire were imported from India and Persia, and diamond from Africa and Brazil. In addition to watch parts, jewels and stones were used in factories as agate lapping dust, diamond chips for drilling, and assorted sapphire tool bits for cutting wheel teeth. One aspect of jeweling which never appeared in American watches was shockproofing. The most common shockproofing system was to mount the balance jewels under slender springs so that the springs would absorb shock rather than pass it to the delicate balance pivots. This had been invented by Breguet in Paris before 1800 but was little used till wristwatches became popular. It might have been a useful idea for pocket watches receiving hard service, such as railroad timekeepers, but apparently was not necessary.

Style of case and type of finish on non-working surfaces of mechanisms were mostly matters of taste and status. Hunting cases were popular in America, more so than in Europe, and lingered well into the 20th Century. They had an opulent appearance and created a self-gratifying ceremony out of telling time, but had no particular function other than perhaps protecting the crystal in a rugged environment. Otherwise, hunting cases simply allowed dirt through their hinges that could impair the watch mechanism, and in any event, their popularity was eroded by the trend for smaller, thinner watches. The final blow was exclusion of hunting cases from the approved railroad watch market since their winding stem was at 3 o'clock rather than 12, presenting a possible source of confusion in reading correct time. In the eyes of the public, railroad

44

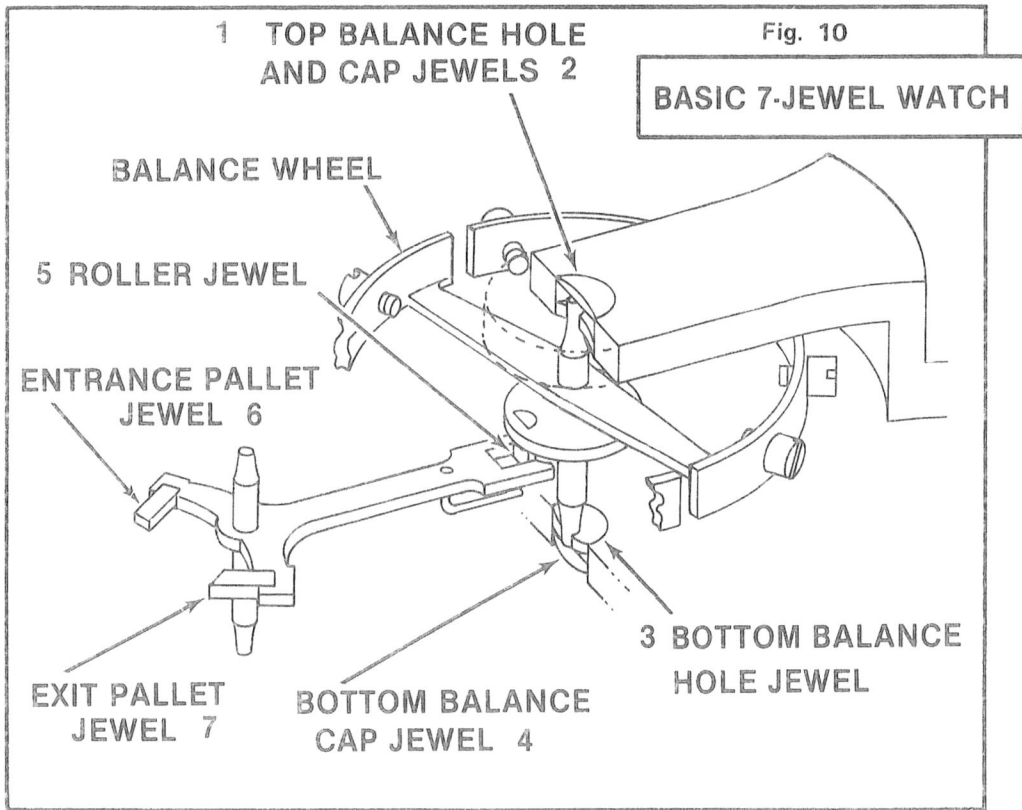

1 TOP BALANCE HOLE AND CAP JEWELS 2

Fig. 10

BASIC 7-JEWEL WATCH

BALANCE WHEEL

5 ROLLER JEWEL

ENTRANCE PALLET JEWEL 6

EXIT PALLET JEWEL 7

BOTTOM BALANCE CAP JEWEL 4

3 BOTTOM BALANCE HOLE JEWEL

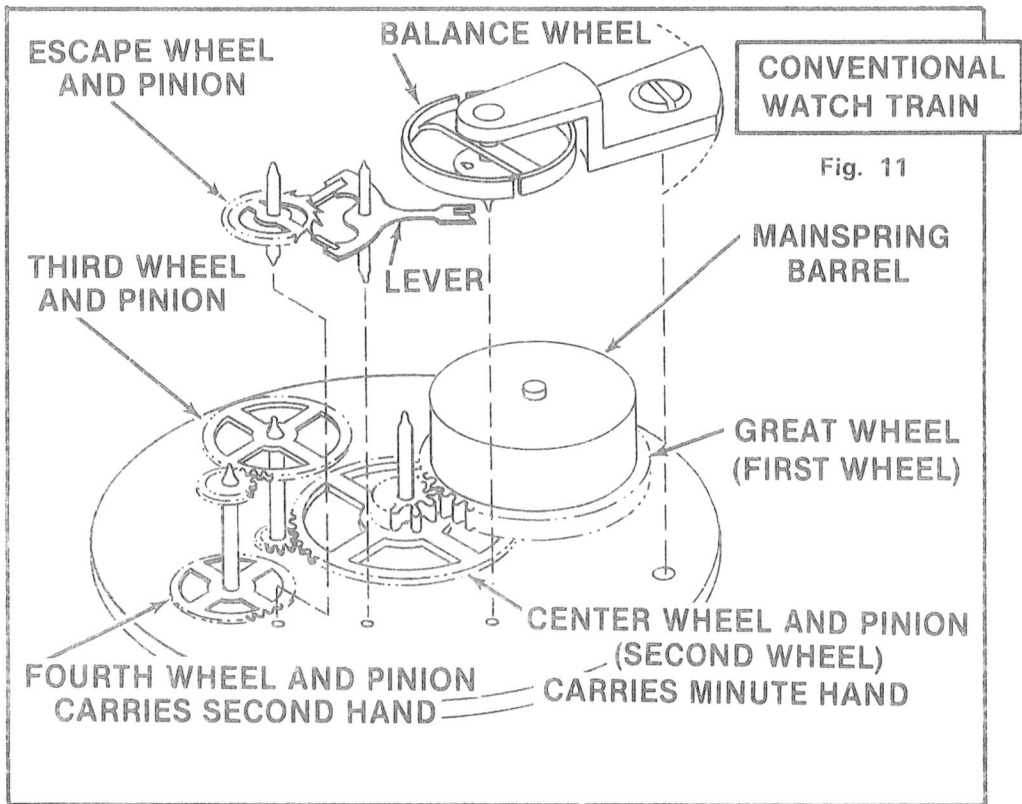

ESCAPE WHEEL AND PINION

BALANCE WHEEL

CONVENTIONAL WATCH TRAIN

Fig. 11

MAINSPRING BARREL

THIRD WHEEL AND PINION

LEVER

GREAT WHEEL (FIRST WHEEL)

FOURTH WHEEL AND PINION CARRIES SECOND HAND

CENTER WHEEL AND PINION (SECOND WHEEL) CARRIES MINUTE HAND

standards defined the one true watch and they wrote the epitaph for the hunting case. Gilded finish on movements had been universal on English watches and was thereby standard on American movements for 40 years. It served well to protect watch plates but required some expertise to create in a fine finish. E. Howard, for instance, long had difficulty obtaining aesthetic r e s u l t s with his gilding. Through the 1850's, however, the Swiss began using nickel finish on high-grade watches. American companies began using nickel finish on finer movements during the late 1860's and on common grades after 1890. Nickel was often plated over brass bridgework although better grades were solid nickel. The Swiss had found that nickel presented an excellent surface on which to buff textured patterns, which they used in several simple designs. Most common were straight line patterns on top plates and radial sunburst designs on pillar plates. Recesses in pillar plates were sometimes embellished with a spotted, or pearled, texture. American companies of course adapted machines for doing this finish work and a delicate geometric style soon developed similar to intricate patterns on Damascus steel, for which it became known as damaskeening. Damaskeening was combined with gold alloy wheels and tall gold jewel settings to produce elaborately decorated watches unique to the American industry. Swiss makers followed along somewhat, using more complicated damaskeening patterns in order to appeal to the American market, while English watches remained gilded.

The mechanism of the average American watch involved the same basic components that had existed for centuries:

1) Mainspring (power source);
2) Geartrain;
3) Escapement and balance (timekeeper).

The power source for watches had always been a coiled mainspring, and in American watches this was usually housed in a going barrel. That is to say that while the early industry modeled their watches after English timepieces they did not incorporate fusees. The fusee, or chain drive, had long been used in European watches to compensate for the reduction in mainspring force as the spring ran down. As verge escapements d i s a p p e a r e d from Continental watches after 1800, French and Swiss makers abandoned fusees for going barrels in order to achieve a thinner design. On the other hand English makers continued with fusees into the 20th Century, for which they achieved a better timekeeper but one that few could afford. Their judgement was theoretically sound and Hamilton used fusees in their marine chronometers during World War II, but such elegance was unnecessary for a common timepiece and even English horology quit the fusee in its last efforts to remain competitive. (Interestingly, a number of English watches with no fusee were made around 1790 using the Debaufre escapement. These came from the area around Ormskirk, north of Liverpool, apparently having little effect on the general English trade.) From the very beginning, American companies used going barrels much like the Swiss, and likewise, it was common to find a Geneva stopwork on the barrel in lieu of a fusee. Use of stopworks was the method by which French and Swiss watchmakers had eliminated the fusee, for if a long thin mainspring was used, a stopwork could limit use of the spring to its middle coils. In this way the extreme inner and outer coils were avoided, where the most dramatic changes in spring force took place, and a fairly uniform driving force was obtained. Even this was no longer used by American factories by 1900, one of the main reasons being that repairmen rarely reset stopworks correctly after cleaning watches, and often left them off altogether.

Going barrels had teeth machined on their periphery, meshing with the geartrain to turn the hands and transmit power to the escapement. Inside was a mainspring, with its outer end attached to the rim of the barrel and its inner end to the barrel arbor, on which the barrel rotated. For winding, the arbor was turned forward, thus coiling the spring inward onto the arbor, and as the watch ran down the spring pulled the barrel forward to catch up. The term "going barrel" differentiated these from both stationary barrels, which were set into the plate, or barrels which were turned for winding, the train then being driven from the center. With a going barrel, breakage of the mainspring could do considerable damage. When the spring broke, its coils rotated rapidly back out against the rim of the barrel with an impact that jerked the barrel sharply backward. This could break teeth off the barrel or center pinion and do damage all the way down to the escapement. To prevent this most American watches incorporated a safety pinion on the center arbor, often called a patent pinion since companies patented their own designs. The barrel meshed with the safety pinion, which was usually threaded onto the center arbor such that mainspring torque held it tight. If the mainspring broke, the sudden reverse shock simply unscrewed the safety pinion without transmitting severe load any further. When the broken mainspring was replaced the safety pinion was screwed back tight, ready to run again.

Most gearing outside horology used the involute tooth form while watches and clocks used epicycloidal gearing. Involute teeth became awkward if gears had fewer than ten teeth, as pinions in watches often had. Also, most machinery used a reducing train, dropping from a high speed input to low speed output since most power sources have been high speed motors and turbines. Quite the reverse, watches required a stepping-up train with an enormous overall gear ratio, which was why pinions of low number leaves were required. This large ratio was apparent, considering that the mainspring barrel made only one rotation in six hours while the escape wheel revolved once every six seconds. This made watches unique and required a highly efficient geartrain, for little power would get through to the escapement if any appreciable losses were present. The losses would also mean that much of the power stored in the mainspring would be wasted in friction, i.e., wear and tear of the geartrain. Watchmakers therefore expended great effort to make well formed gearteeth for friction-free operation, and used small diameter pivots running in lightly oiled jewel bearings. Jewels reduced friction and resisted wear, since they were quite hard, thus maintaining wheels in their correct locations. Where jewels were not used the pivots were slightly larger to reduce loading and wear. It was obviously important to develop machinery for cutting p r e c i s i o n teeth. Traditional European craftsmen had formed wheel and pinion cutters by eye, with results ranging from poor to excellent depending on the patience and skill of the workman. Early in the American industry Charles Vander Woerd developed his machine for forming correct cutters. This enabled production of consistently fine wheels and pinions, also allowing use of weaker mainsprings since more of the energy was sure to reach the escapement. Weaker mainsprings in turn reduced forces throughout the geartrain along with associated friction and wear. So it can be seen that machine capability could mean consistent quality and better watches.

In watches, wheels always drove smaller pinions and the pinions, except for safety pinions, were cut directly into steel arbors, while wheels were usually of brass or gold alloy and pressed onto the arbors. Steel pinions were polished brightly to reduce friction, and hardened to resist wear. It might be thought that soft brass wheels would wear faster than hardened pinions but that was not the

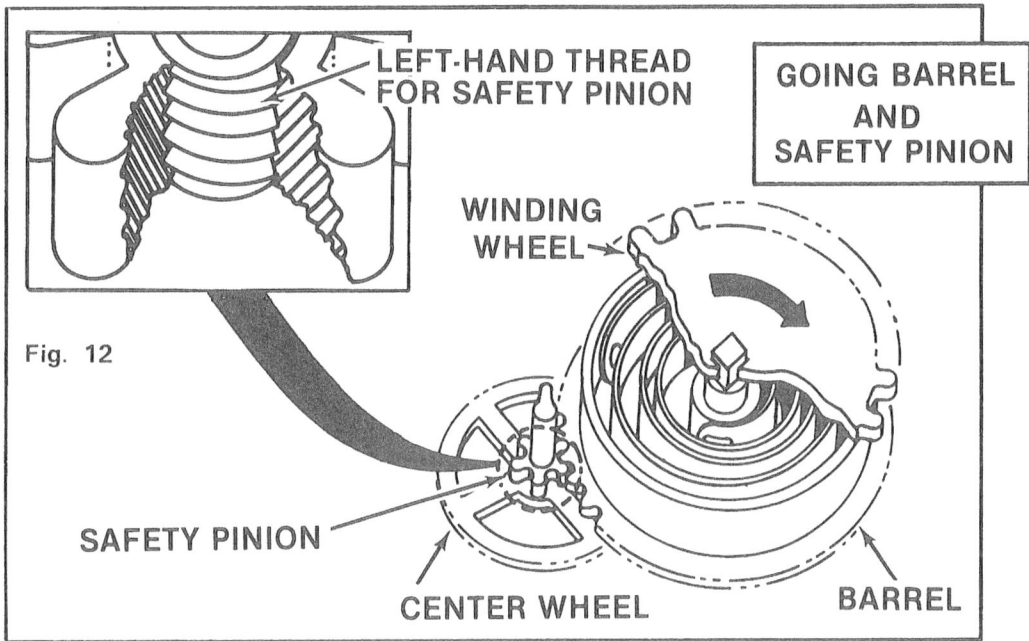

LEFT-HAND THREAD FOR SAFETY PINION

GOING BARREL AND SAFETY PINION

Fig. 12

WINDING WHEEL

SAFETY PINION

CENTER WHEEL

BARREL

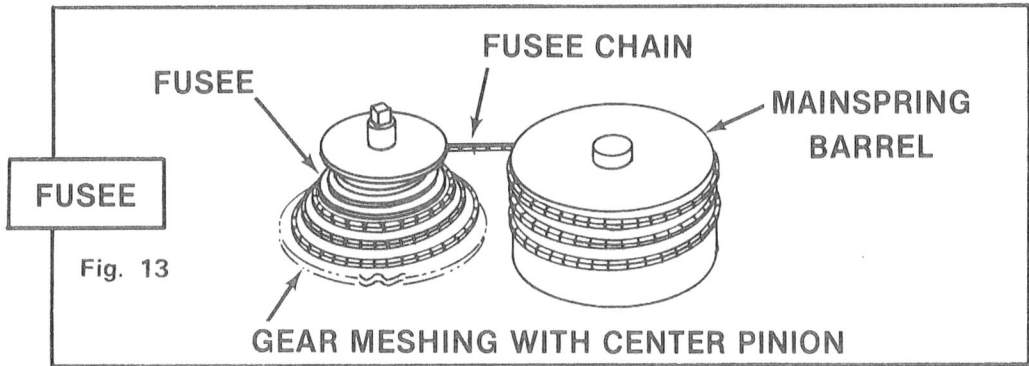

FUSEE

FUSEE CHAIN

MAINSPRING BARREL

FUSEE

Fig. 13

GEAR MESHING WITH CENTER PINION

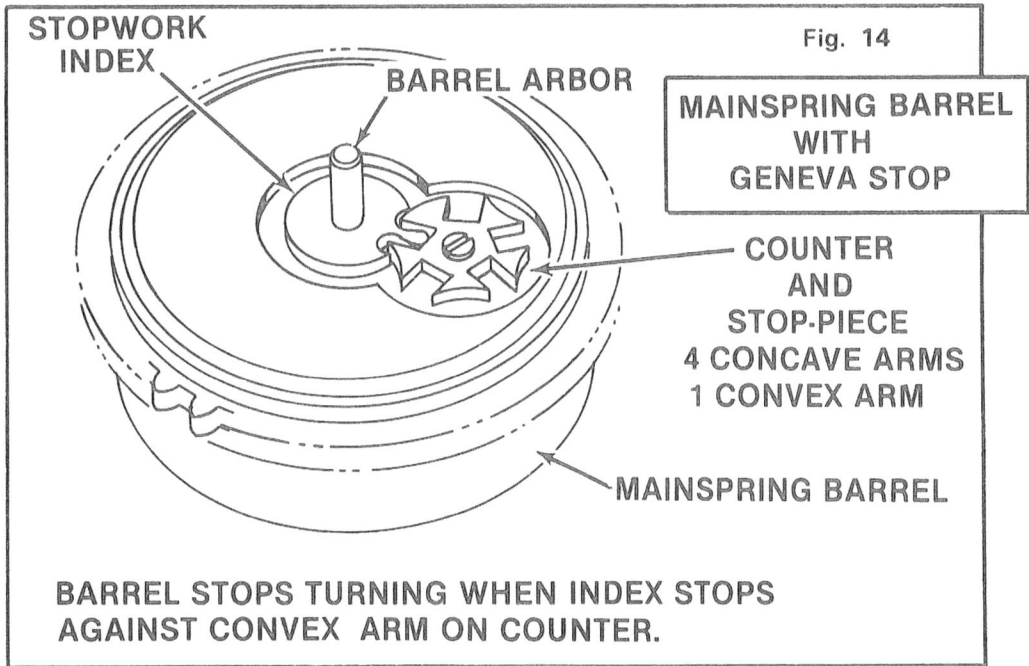

STOPWORK INDEX

Fig. 14

BARREL ARBOR

MAINSPRING BARREL WITH GENEVA STOP

COUNTER AND STOP-PIECE 4 CONCAVE ARMS 1 CONVEX ARM

MAINSPRING BARREL

BARREL STOPS TURNING WHEN INDEX STOPS AGAINST CONVEX ARM ON COUNTER.

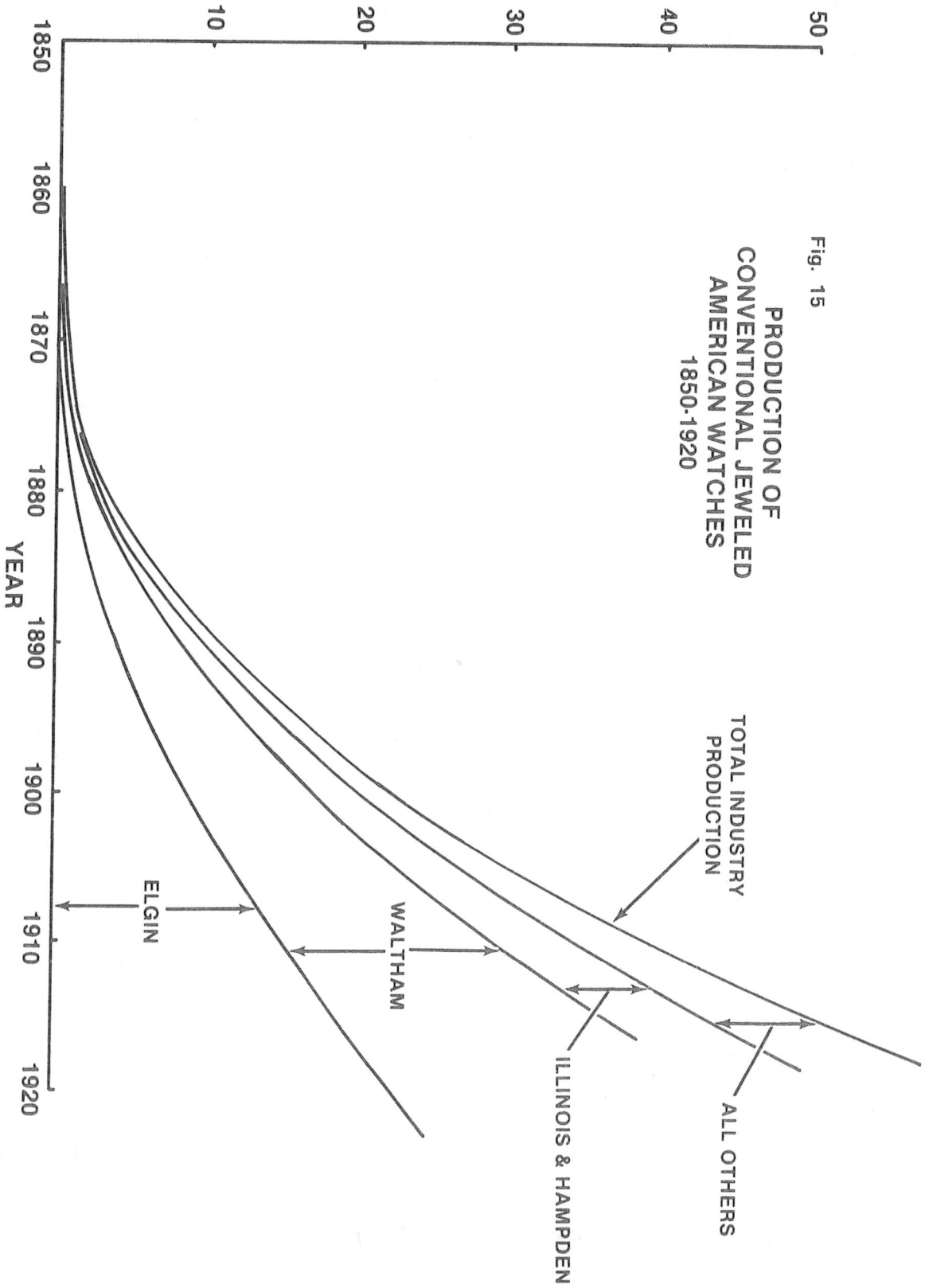

WATCH PRODUCTION · MILLIONS TO DATE

Fig. 15

PRODUCTION OF
CONVENTIONAL JEWELED
AMERICAN WATCHES
1850-1920

TOTAL INDUSTRY
PRODUCTION

ELGIN

WALTHAM

ILLINOIS & HAMPDEN

ALL OTHERS

YEAR

1981 © M. C. HARROLD

48

case. Inevitably, tiny bits of dirt became imbedded in the softer wheel teeth and these hard particles wore the steel pinions, severely notching them if allowed to continue. Therefore, frequent cleaning was essential for high-grade timekeepers in hard use, such as railroad watches.

Detached lever escapements were universal in American factory watches, except for some dollar watches. Even in 1838 the Pitkins used this escapement since it was then gaining respectability in Europe. In addition, it was more readily producible than cylinder escapements being used in many Swiss watches and offered advantages of being a good timekeeper, durable, and easy to repair. Its timekeeping ability rested with the fact that it was detached, leaving the balance wheel free during most of its operating sequence, whereas cylinder and duplex escapements left their escape wheel locked against the moving balance most of the time. Lever escapements interfered less with proper oscillation of the balance and hairspring, which was at the heart of good timekeeping. The term "detached" lever was used to differentiate these from rack levers which were popular for a time in England, and not detached at all. Modern books still specifically refer to "detached" levers and considering that rack levers died away around 1825 it's surprising how long traditional terminology can persist.

For temperature compensated watches, the balance wheel was made bimetallic with steel inside the rim and brass on the outside. English chronometer watches had been temperature compensated as early as 1780 to correct for thermal effects on hairsprings, but common timekeepers weren't accurate enough to warrant compensation till after 1840. American factory watches were not generally temperature compensated till the late 1850's and then only in better models. By 1880 most grades of jeweled movements had bimetallic balances as the quality of average watches improved.

Needless to say, the average American watch was made in the average American watch factory. Excluding dollar watch companies, this would have been an organization of perhaps 200 people producing 150 watches per day. There would have been separate departments making plates, trains, flat steel, escapements, balances, and motion work, plus departments for jeweling, gilding, finishing, and adjusting. Otherwise it was likely purchasing jewels, mainsprings, dials, and perhaps whole balances from outside sources, at least for the first few years. It was probably less than ten years old and was soon to go out of business, for watch companies came and went rather rapidly. For some reason many people were inclined to open watch factories during the last half of the 19th Century, even though few managed to accumulate long histories. A number of these were organized out of remnants of a defunct company and consequently produced an extremely similar watch to their predecessor's. Many others were independent new entities making watches of their own design. This thankful proliferation was one of the few forces bringing variety to the industry, and from small companies came some of the most interesting specimens of American watchmaking.

On the other hand, one could insist that the average watch obviously came from Waltham or Elgin, for on any given day half the jeweled movements manufactured in America came from these two factories. The industry had produced 25 million jeweled timepieces by the turn of the century, with 9 million coming from Waltham and 9 million more from Elgin. The situation was more extreme after 1900 since there was less competition than in earlier years and the Big Two had an even greater percentage of total sales. Both these factories employed 2500 people manufacturing 1200 watches per day, and since millions of

their watches were spread across the continent, they had a lucrative business in spare parts. Of course companies with that share of the market were extremely influential in setting styles and prices. They were the competition that everybody else had to match, with the finest and most efficient equipment and many years of experience. Uniformity in the watch industry tended to revolve around these two companies, not just because they could influence the tenor of business, but because much of the industry naturally evolved from them. Since many new firms opened their shops with designers from Waltham and Elgin, watches throughout the industry were permeated with traits of the two giants. Be that as it may, it was the myriad of little companies that painted the overall picture of American watchmaking. The last 30 years of the century averaged one new watch company per year, with a death rate not far behind. Looking back we see a blur of names flashing past, some so brief there was hardly time to produce anything. American watchmaking is remembered as a boom industry characterized by a rapid flurry of activity in the late 19th Century. Scarcely a dozen companies survived 40 years while 40 companies lasted less than ten and many never reached age five.

After 1880 the Average American watch began assuming a new identity, the "dollar watch." Till then factories had been making by machine a product that was much like traditional watches constructed by hand. Substantial plates were machined on lathes, set with jeweled bearings, and placed over quality wheelwork and escapements to complete durable timekeepers adjusted for accuracy. Much of the population could not afford such durability and precision nor did they need it. They were willing to buy some kind of watch however, and the era of competition saw a new kind of pocket watch arrive that eventually (1896) dropped to a retail price of one dollar. Its design was based on minimizing the number of parts while maximizing use of economical punch-press methods, and by the turn of the century this practical device matched sales volume of conventional jeweled watches. Several companies tried upgrading the dollar watch recipe to an inexpensive jeweled watch, but usually found that conventional watch factories could out-compete them. Full-blown dollar watches were another story, with manufacturing and marketing systems few conventional makers entered. They were less demanding to make, but in great demand, so production quantities increased rapidly. This by no means ended conventional watch manufacturing. Several important new companies entered the jeweled watch market in subsequent years, such as Hamilton, Keystone Howard, and South Bend, but the pace of jeweled watch manufacture leveled and after 1910 began to decrease. Once dollar watch making established a broad manufacturing base it remained the dominant force in United States watchmaking.

Table V
Total Production Quantities 1900
Conventional Jeweled Watches

Waltham	9,000,000
Elgin	9,000,000
Hampden	1,300,000
Illinois	1,000,000
Seth Thomas	500,000
Rockford	410,000
Columbus	300,000
U. S. Waltham	225,000
Lancaster	150,000
Aurora	150,000
Peoria	100,000
Howard	100,000
Hamilton	100,000

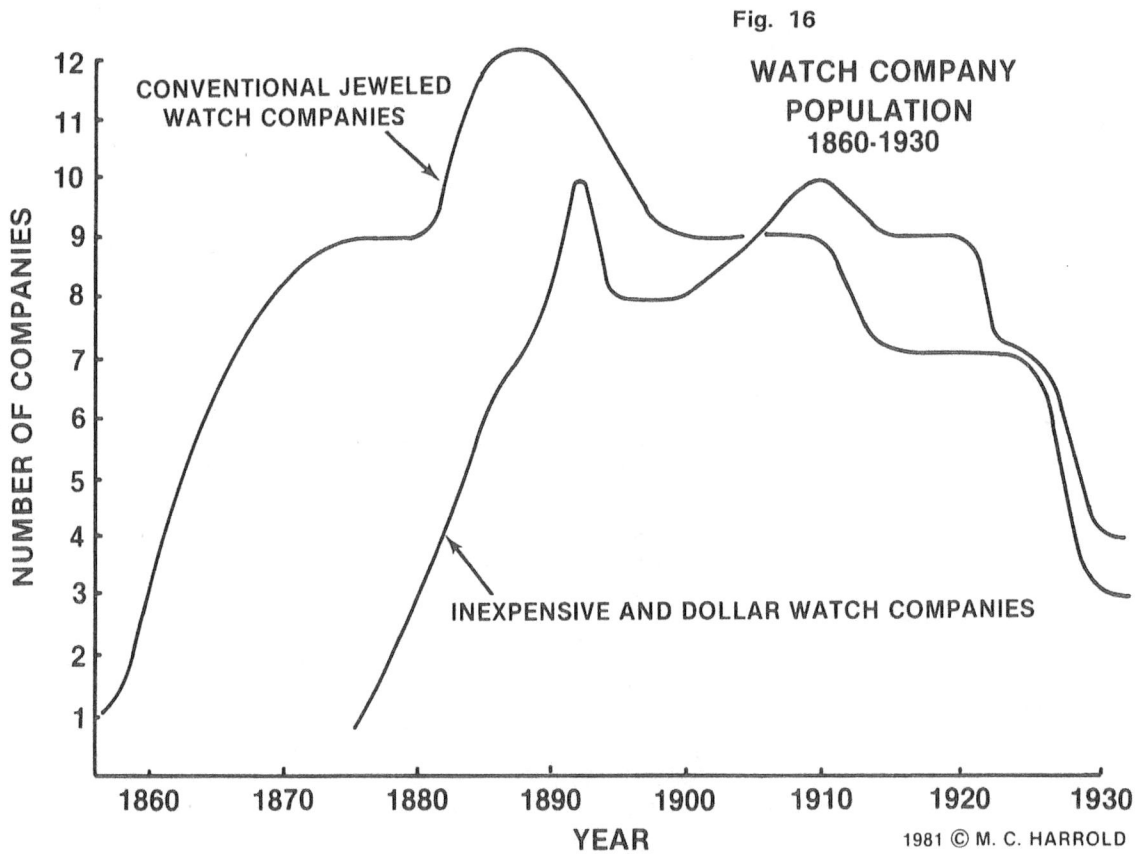

Fig. 16

WATCH COMPANY
POPULATION
1860-1930

CONVENTIONAL JEWELED
WATCH COMPANIES

INEXPENSIVE AND DOLLAR WATCH COMPANIES

NUMBER OF COMPANIES

YEAR

1981 © M. C. HARROLD

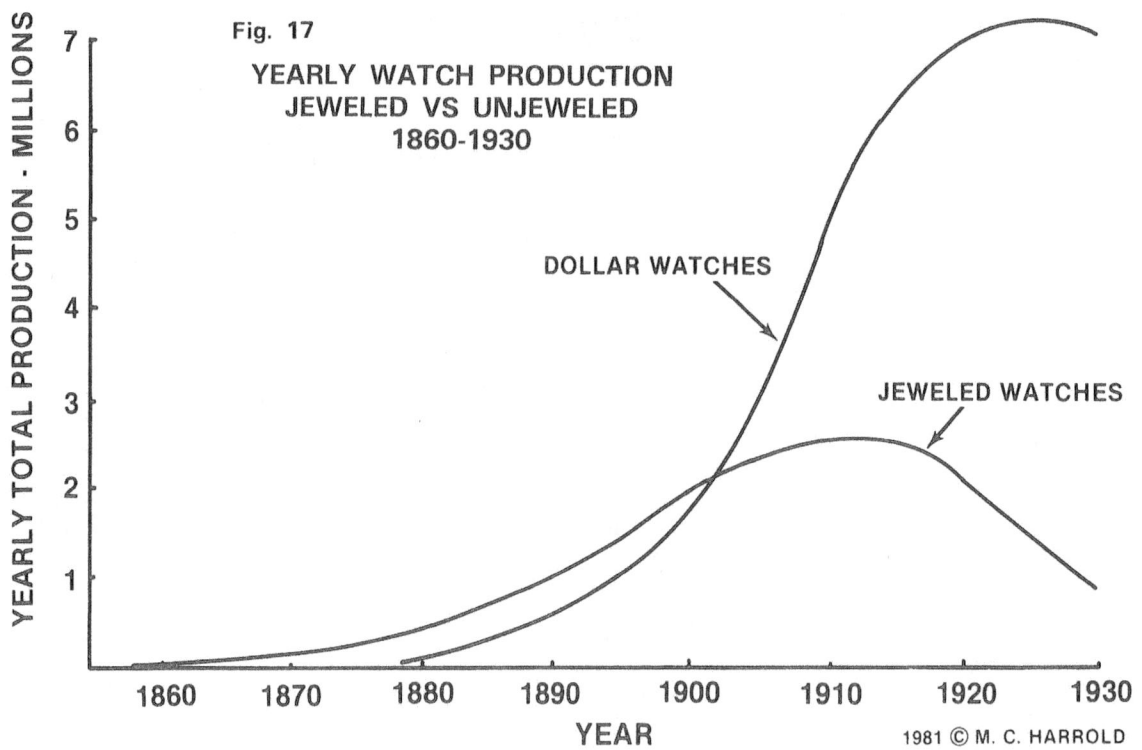

Fig. 17

YEARLY WATCH PRODUCTION
JEWELED VS UNJEWELED
1860-1930

DOLLAR WATCHES

JEWELED WATCHES

YEARLY TOTAL PRODUCTION - MILLIONS

YEAR

1981 © M. C. HARROLD

New York Standard	1,200,000
Trenton	600,000
Manhattan	500,000
Knickerbocker	500,000
Cheshire	100,000
Suffolk	30,000

Dollar Watches

Waterbury Watch Co.	7,500,000
Ingersoll	6,000,000
New Haven Clock Co.	5,000,000

While cheap timepieces were thriving, jeweled watchmaking still had glory to come. Railroad watch standards, issued in 1892, drew attention to quality jeweled timekeepers, and conveniently prescribed what they should be. The activity that followed generated new models, new companies, increased demand, and some of the watches most remembered today. Accurate watches were already connected with the romance of the iron horse when American watchmaking began in the 1850's, and American companies started filling specific railroad orders in the 1860's. For the most part, however, these watches had no readily identifiable features, being more distinguishable by performance than appearance. Railroad watch standards of the 1890's specified details that could be seen, plus standards of quality and performance that defined railroad watches as being above-average timekeepers and as being American made. Railroad watches thereafter became discernable prestige items, so that models, grades, and sizes not suitable for rail service often adopted popular railroad styles. The railroad market gave companies an opportunity to mass-produce reputable watches in fine fashions that factories well knew how to make. The most widespread effect of this was to popularize 16-size movements and split-plate construction, the latter of which spread to most other sizes as well. This did not really become an overshadowing force, for instance it could not stem demand for small 12-size watches which the standards did not allow. It certainly did create a new demand in a period when growth of jeweled watch sales was falling, constituting a vitalizing influence in the declining years of pocket watches.

Dollar watches and railroad watches fueled the competitive fires as companies scrambled for their share of sales. Years of expansion had created more companies than markets could easily support, each factory having sophisticated machinery capable of mass-production. The average American watch was thus produced in huge numbers, at an ever decreasing price, with a correspondingly decreasing profit margin. This seemingly vast flow of wealth brought a growing intensity to watch manufacture, requiring companies to maintain high production while seeking a perilous financial balance between competitive austerity and stable reinvestment. Recessions in the 1890's therefore created instability in the trade and many companies foundered. Few surviving firms were financially solid, so that American watchmaking decayed after the 20th Century began rather than partake in the economic recovery enjoyed by most of the nation. Over-competition had decimated the watch industry.

Ill health in American watchmaking began exhibiting several symptoms after 1900: traditional markets approached saturation and there was a lack of new companies. Beginning with jeweled watches, Figure 17 shows that while production peaked in 1910, it had been decelerating since the mid 1890's; meanwhile, the industry was too debilitated to build new business as production approached zero growth. Jeweled watchmaking had already reached zero population growth by 1900, as far as number of companies was concerned, which can be seen in Figure 16. The only new companies formed after 1900 were South Bend (1903) and Manistee (1908), which were more than offset by the deaths of other companies. Contributing to decay of the jeweled watch industry was massive popularity of dollar watches. Figure 17 demonstrates that growth of dollar watch production did not begin rounding off till 1910. Bannatyne Watch Company, formed in 1905, was the only new organization attracted to dollar watch making after 1900. Other entrees to the field were existing clock companies seeking new sources of income: Westclox (1900), Ansonia (1904), and E. Ingraham, which bought Bannatyne in 1911. These firms were utilizing existing manufacturing facilities, distribution systems, and company reputations to increase revenues. This sort of consolidation also occurred among jeweled watch companies as Keystone bought Howard (1903), Ingersoll absorbed Trenton (1907) and New England (1912), and Waterbury Clock purchased Ingersoll (1922). After the turn of the century, the last few watchmaking entrepreneurs took their chances, vitality was gone, and operating reserves were dwindling. Aging management organizations, accustomed to boom times that had established their reputations, could not accept that the weather had changed or that the Swiss were still serious challengers.

The Swiss were serious indeed. They had invested in machine technology, updated their marketing techniques, and developed a new product: the wristwatch. Through their perseverance, wristwatches became the final form of the average American watch. Small watches, some in finger rings and bracelets, had occasionally been made during the 18th and 19th Centuries, but were rarities. Then, around 1880, Girard Perregaux and several other Swiss firms received commissions from Berlin to manufacture wristwatches for naval officers. (While these were presumably intended to increase combat efficiency, they were in gold cases.) The Swiss began selling similar watches commercially and by the turn of the century some were coming to America.[4] A number of those were also being returned to Switzerland as unacceptable for American taste. The Swiss persisted, displaying at horological and jewelry exhibits, so that bracelet watches became part of Paris fashion. Fashion appeal popularized wristwatches among women and World War I introduced men to wristwatches. Modern warfare had become a synchronized exercise, and bracelet watches were less encumbering than pocket timepieces. Commercial men's wristwatches after the war generally utilized a military style rather than risk an effeminate appearance. With a more frivolous attitude during the 1920's, men's wristwatches grew vastly more popular, a trend also encouraged by an increasingly urbane American self-image as the wild west finally became tame. Wristwatches were a logical convenience throughout Europe and America by 1930, having completed a half-century of struggle to become accepted.

For American watchmaking, the wristwatch was a rising flag on a sinking ship. Companies first used 0 size ladies' movements which required no new tooling. Illinois and Elgin introduced ladies' wristwatches around 1907. Gruen first carried bracelet watches in 1908, and with Swiss manufacturing facilities, they were probably better equipped to pursue the wristwatch market. Waltham's 1911 catalogue listed no bracelet watches at all. By 1914, numerous ladies' wristwatches and a few men's strap watches were being marketed by Waltham and other American companies.[5] World War I boosted production, allowing some firms to make new tooling at government expense, but the Roaring 20's created a dilemma as wristwatches became extremely popular. Styles were growing to diverse sizes and shapes, each requiring new tools, and large quantities of machinery were required to fabricate correspondingly large quantities of watches. Aged and ailing watch companies therefore faced the decision of financing new automated production lines or going out of business, the latter of which proved more attractive to many, such as Seth Thomas, Illinois, Keystone Howard, and New York Standard. Companies did not have either capital or borrowing power to update their factories, so that depression hit the watch industry even before the crash of 1929.

By 1930, the spring had run down on American watchmaking. The seven companies that survived the 1920's also survived the Great Depression:

Jeweled Watchmakers	*Dollar Watchmakers*
1) Waltham	1) New Haven Clock
2) Elgin	2) Waterbury Clock
3) Hamilton	3) Westclox
	4) E. Ingraham Clock

They enjoyed strong historical reputations so that traditional buying patterns supported a moderate American watch industry for another generation. These companies operated from a weak financial position, however, and an eroding share of total watch sales. Younger generations were being reared in a world that accepted Swiss watchmaking as the modern trend. The only remnants we see today of those seven survivors are Westclox and Timex, which organized out of the Waterbury Clock Company. The dollar watch, a yankee invention based on American industrial methods, was proven one of the most durable contributions of America's great watchmaking era.

Certainly the most enduring creation of American watchmaking was its concept of machine-made interchangeable parts. When the industry began, in 1850, guns and clocks were the few mechanical contrivances owned by consumers, and those had become commonly available only to the previous generation. Complicated demands of watchmaking resulted in increasingly sophisticated machinery which, in turn, made possible more sophisticated and plentiful consumer products. Machines were just then growing from occasional industrial tools to servants of everyday life. Whereas machines rarely entered private homes before the Civil War, years following saw a proliferation of household gadgets ranging from apple peelers to sewing machines and typewriters. Manufacture of such complicated conveniences required not only intricate machine tools, but automatic ones that produced large quantities of goods. This economy of scale resulted not only in low cost goods, but also high volume so that gross sales revenues could more than pay for the expensive tools. It was this economic cycle that created the modern standard of living. It was the precision required for watch manufacturing that generated tools capable of meeting modern technical requirements.

REFERENCES

1. Exploits of these individuals have been taken from: Charles S. Crossman, *The Complete History of Watchmaking in America* (reprinted Exeter, New Hampshire, Adams Brown Co.).
2. Graphs of jeweled watch trends were obtained by sampling data from serial number lists of Waltham and Elgin.

These two companies were studied since they represented such a large share of the jeweled watch market and manufactured watches of all types. The other large companies (Illinois, Hampden, and Seth Thomas) had a similar range of products and would not significantly alter the results. Smaller companies were not large enough producers to change the average trends. Also, small companies often concentrated in a particular part of the market, as with Hamilton.

3. Jaquet & Chapuis, *Technique and History of the Swiss Watch* (Olten, Switzerland, 1953), p. 180.
4. Ibid., p. 116.
5. Anthony Gohl, *The Wristwatch* (BULLETIN, The National Association of Watch and Clock Collectors, Inc., December, 1977), p. 586.

Table VI

Total Production Quantities 1930

Conventional Jeweled Watches

Elgin	33,000,000
Waltham	27,000,000
Illinois	5,300,000
Hampden	4,600,000
Seth Thomas	3,500,000
Hamilton	2,500,000
Columbus & South Bend	1,200,000
Rockford	1,000,000
U. S. Waltham	450,000
Howard	250,000*
Aurora	150,000
Lancaster	150,000
Peoria	100,000
U. S. Marion	20,000
Newark	13,000
TOTAL	**79,233,000**

Inexpensive Jeweled Watches

New York Standard	3,000,000
Knickerbocker	2,000,000
Trenton	1,600,000**
Manhattan	500,000
Cheshire	100,000
TOTAL	**7,200,000**

Dollar Watches

Ingersoll	74,000,000
Westclox	50,000,000
New Haven Clock Co.	30,000,000
Waterbury Watch Co.	12,000,000
Ansonia Clock Co.	10,000,000
E. Ingraham	10,000,000
TOTAL	**186,000,000**

*Keystone Howard production is estimated at 150,000
**Includes 350,000 Ingersoll Trenton movements

V
INEXPENSIVE WATCHES

The first watches by the Warren Mfg. Co. sold for about $40 in 1853,[1] no small price at a time when the average worker took home perhaps a dollar a day. Few could afford them and, indeed, there was only limited need for a person to have a watch in that era. The town clock kept most people adequately informed of the time. By then a person was also likely to have his own clock at home, possibly with strike and alarm. The development of mass-produced Connecticut shelf clocks in the 1830's had brought the price below $10 and made them available in many areas through local drummers who sold them on time payments, pun intended. So a simple clock was something an average person could better afford. This was symptomatic of an ever growing awareness of time throughout the second half of the 19th Century. As the country became industrial rather than agricultural there was more need for people to keep track of the business day since its schedule was increasingly a part of their lives. Even if they were not directly involved in the industrialized society, there was no way to ignore the whistle of the trains as the railroads spread their network across the land. For better or worse the whole country was beginning to operate on a timetable, just like the railroads, and imperceptibly time became a part of everybody's lives. As these forces began building a demand for watches, which supported the rise of the industry, they built a demand for watches which everybody could afford, and which the early industry was not able to produce. People wanted a watch which was inexpensive but a reliable and useful timekeeper, able to last at least several years and capable of being repaired.

The first American company to begin producing low-cost watches was, naturally enough, the first company. The Boston Watch Co. had manufactured its C. T. Parker as a low-grade 7-jewel watch, which may be guessed to have sold for under $20. Boston's successors, Appleton, Tracy & Co., continued with a low-cost 7-jewel movement marked "P. S. Bartlett," an old Boston Watch Co. model name. In 1861 the firm, then the American Watch Company, introduced a variation called the J. Watson, which later that year was renamed "William Ellery," who had been a signer of the Declaration of Independence. All along, these names

75. William Ellery #67,234 KW18 American Watch Co. ca. 1862.

had applied to virtually identical versions of the Boston Watch Company's 18-size fullplate, since known as the model 57. As the William Ellery, it coincided with increased demand spurred by the Civil War, when sales moved briskly. The Ellery grade was also manufactured in the KW18, and after Waltham's purchase of Nashua the KW18 was used as a low-cost model till being discontinued in 1868. In that year, Waltham introduced a cheaper fullplate movement to compete with products of newly formed American companies, as well as inexpensive Roskopf watches which began arriving from Switzerland. This model was marked simply "Home Watch Co. — Boston," possibly in an attempt to keep the Waltham reputation separate from the cheapest grade of goods. Using the same marketing logic, U.S. Marion responded with their "North Star — New York" in 1871, but it was never as successful as the Waltham model. The "Home Watch Co." sold well until being replaced in 1873 by the even cheaper "Broadway," which probably retailed at $10 or less as a cased watch.[2] These inexpensive grades were usually 7 jewels, but occasionally had their top plates jeweled, bringing them to 11 jewels. Still, there was the competition of Swiss cylinder watches which sold for as little as $8. The cheap Swiss watches were churned out by hand-work methods, with subdivision of labor as the means of making them inexpensive. Each step of the manufacturing process was performed rather hastily by hand, with the result that quality varied and was generally low. It followed that the watches were as erratic in their timekeeping as in their mechanical perfection, often requiring repair and correction when they were still new. This did little to enhance the reputation of Swiss watches, which were certainly not all of this low order. It also pointed out that machine manufacture could overcome the major barriers to good cheap watches with production of uniform quality and elimination of hand finishing.

While the number of parts was held to a minimum, the Broadway grade was not really a new design, it was merely manufactured in a manner that was convenient for the American Watch Company. The first effort to design a truly good cheap watch did not occur in the United States but

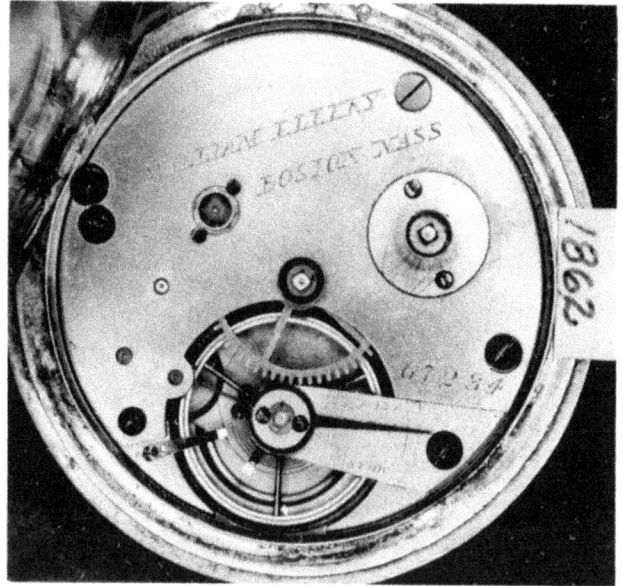

74. C. T. Parker #1171 Boston Watch Co. ca. 1854.

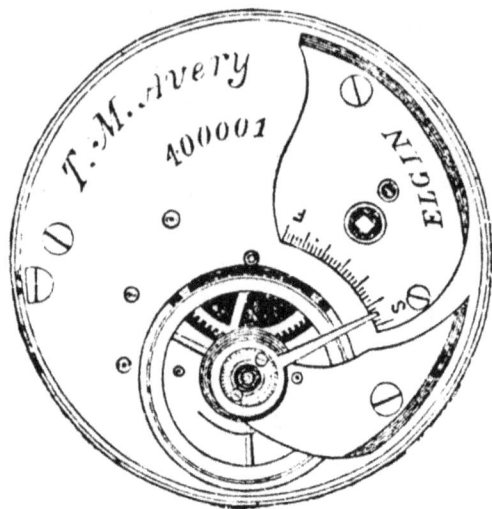

76. Elgin T. M. Avery — inexpensive model ca. 1869.

was the work of George Roskopf (1813-1889) in Switzerland. Roskopf was an idealist who wanted to make a useful watch available to the humblest person, and at the same time boost the Swiss industry. Besides avoiding any luxury and finish work, he designed a movement so as to have fewer parts, an escapement on a detachable unit so it could be adjusted separately, a plain nickel case with no hinge on the back, and hands that were set by turning them with a finger, as on a clock.[3] He also used the pin pallet escapement invented in 1798 by Louis Perron but little used till Roskopf took it up.[4] This escapement used upright steel pins for pallets, rather than jewels, and the lever banked against the rim of the escape wheel. (Today the pin pallet escapement is universally used for inexpensive watches and alarm clocks.) He was able to market the watch at a price of perhaps $6 in 1868 and a model with stem-setting hands in 1870. To produce these he had to fight an uphill battle against the traditions of the mountain craft system in Switzerland. Though it was not a machine-made watch, his product and his methods were not those that the craftsmen knew or liked and it was only by determined effort that he was able to get work done. His perseverance paid off, for the watch became a success. Wishing to see the benefits go to the Swiss industry, he patented his designs in a number of countries around the world, except Switzerland, and Swiss factories were soon exporting vast numbers of Roskopf-style watches, many to America.

As cheap Roskopf watches were becoming popular, Elgin and Waltham were both marketing inexpensive-grade fullplate movements, while Newark and U. S. Marion largely avoided the competition for low-price sales and were making mostly 15-jewel watches. Elgin then introduced other inexpensive watches named Leader and T. M. Avery. These were 17-size fullplate movements, both stem and key wind, which Elgin sold cased. At perhaps $5 or $6, these were still expensive and effort continued to make a truly low-cost watch. Among the people searching for a cheap watch design in this country were two patent solicitors, Edward Locke of Boston and George Merritt of Brooklyn. Since they were in the market for clever ideas, the rotary watch of Jason R. Hopkins was brought to their attention. Hopkins, of Washington, DC, had been a founder of the short-

lived Washington Watch Co. which produced some tools about 1875 and material for about 50 keywind movements that were ¾ plate with duplex escapement.[5] Around this same time he invented an unusual watch with detent escapement and a rotary movement which he hoped could be sold for as little as 50¢.[6] The rotary movement was an arrangement by which the works of the watch rotated around inside the case approximately once every 2½ hours, driven by a long mainspring. While this may sound complex, it could be accomplished with fewer parts than a normal style watch, therefore having the potential of being cheaper to make. Hopkins showed Mr. Locke a model watch in 1867 but Locke was not initially enthusiastic over it. Nor was the firm of Benedict & Burnham in Waterbury, Connecticut, a maker of stamped brass parts for the Connecticut clock industry, in which Locke hoped to make his cheap watch. But George Merritt encouraged Hopkins to improve the design and after some changes it was again inspected. The entrepreneurs were still not satisfied and gave the Hopkins rotary watch no further consideration.

Still hoping for success, Jason Hopkins joined with Mr. William A. Wales of New York, who had been an organizer of the United States Watch Co. of Marion, New Jersey. Together they approached William B. Fowle of Auburndale, Massachusetts, who purchased the design for $10,000 and set up a company for manufacturing it.[7] Machinery was purchased from the assignee of the United States Watch Co. even though it was not really suited for the job, a shop established in Auburndale, and movements produced in 1877, selling for $10 wholesale. While the model had been detent escapement, production watches were lever escapement with stem wind and set, and 18-size cases. Generally they did not run well and many were returned to the makers, probably not so much due to failure of the overall design as to poor construction of the details. The scheme was to put the large mainspring in an equally large barrel, then set the train and bridge-work on top the barrel using the barrel as a pillar plate. As the movement rotated with the barrel, the geartrain was driven by meshing with a stationary internal gear planted around the movement. This was a workable idea but probably required more pre-

77. Auburndale rotary #99 Auburndale Watch Co. ca. 1877.

78. Auburndale Bently #2 ca. 1880.

cision in manufacture than a cheap watch warranted, or at least more than the Auburndale group was able to produce on the machinery they used. The enterprise had begun on the basis that a large production rate could be achieved with little capital. This was not possible, as Aaron Dennison had long since learned, and the result was an inadequate timepiece, even for an inexpensive watch. So after about 1000 had been made, few of which had been sold, the rotary watch was abandoned. The company switched to manufacture of a timer, that is, a watch not intended for telling the time-of-day at all, but a stop watch for timing intervals such as horse races. With production of the Auburndale Timers going somewhat successfully, the company tried again to market inexpensive watches. These were more conventional 18-size movements, ¾ plate with gilt finish, designed to fit regular American hunting cases. There were two models named after the sons of Mr. Fowle, the "Lincoln" being keywind and the "Bentley" stemwind. Unfortunately it was again found that these could not be manufactured economically and they were given up after only several hundred were made. While the Timers continued to carry the company, Auburndale next went into production of metallic dial thermometers in a variety of sizes, which were successful and constituted the main product of the firm during its last years. The Timers were dropped in 1881 since sales were sporadic and a large inventory had accumulated. By 1883 more capital was required than could be found, forcing the company to close. All along, the main source of funds had been the personal fortune of Mr. Fowle, but the misadventures of the company consumed this and the gentleman went bankrupt along with the Auburndale Watch Company. In its brief life the firm made several interesting watches, none in any great quantity, and was the first American company to attempt manufacture of either a cheap watch or a genuinely inexpensive jeweled watch. As with the Pitkins' unsuccessful

venture in 1838, Auburndale's futile effort was an indicator that the time and technology were ready for major changes in watch manufacture.

Merritt and Locke, who had wisely rejected Hopkins' rotary timekeeper, had the determination to make cheap watches happen. While walking a Worcester, Massachusetts, street in 1877, Edward Locke noticed a tiny steam engine in the window of a watch repair shop, and sensing at least an interesting story, went in to investigate. The engine had been built by the proprietor of the shop, Mr. Daniel Azro Ashley Buck, and furthermore had been exhibited at the Centennial Exposition in Philadelphia, next to the huge Corliss steam engine which powered the exhibition halls. The little engine was only ⅝ of an inch tall by ⅛ inch square and composed of 148 parts held together by 52 screws. Locke felt that the designer of such a marvel might well be the man who could invent the watch he was looking for, and an agreement was reached on the spot; Mr. Buck to make a model watch for $100. When the model arrived for approval it was a rotary watch with long mainspring and detent escapement, similar enough to the Hopkins watch that Locke and M e r r i t t were less than enthusiastic. Buck became ill shortly thereafter but had taken note of their objections and, while convalescing, conceived ideas for improving the watch. In the fall of 1877 a new model was complete, with a new plate layout and Buck's own version of the duplex escapement, which would be quite inexpensive to manufacture. This appeared much more promising so that the trio of Locke, Merritt and Buck journeyed to Waterbury to see if the Benedict & Burnham Mfg. Co. would be interested in this design. Upon inspecting it Mr. Benedict was satisfied and set apart an unused portion of the shop in which to make the watch. Only $8000 was ventured on the enterprise, second-rate machinery was used, and they had to educate employees. They still managed to produce 150 watches per day, getting to market in late 1878 and selling watches as fast as they were ready. The experiment was an instant success, so Merritt and Locke could see they were right in predicting a market for cheap watches.[8]

Success of the design was based on the punch press, the most economical method of machine manufacture, plus ingeniously simple construction. Locke's and Merritt's reasoning was clearly directed toward a stamped brass watch. Their rejection of Hopkins' design, their selection of Benedict & Burnham as a manufacturer and the Locke/Merritt /Benedict approval of Buck's model was all based on ability to punch Buck's longwind watch from sheet brass. Stamped brass construction had long since dominated the Connect-

THE WATERBURY FACTORY.

79. Factory of the Waterbury Watch Co. ca. 1885.

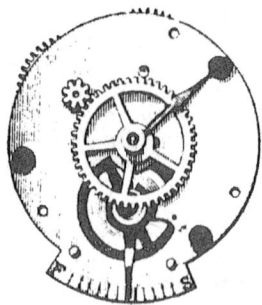

The hour wheel here shown, with the hour-hand attached, turns on the same pinion as the center-wheel, but having a less number of teeth, skips every other tooth of the pinion, hence goes slower as required. The regulator also here shown shortens or lengthens the hair-spring, making the watch go faster or slower.

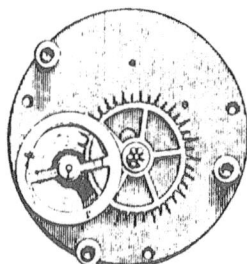

Showing Duplex Escapement of the Waterbury.

THE power being applied direct (see page 3), the Waterbury dispenses with the full train of wheels in ordinary Short Wind Watches, there being only two wheels in its train as here

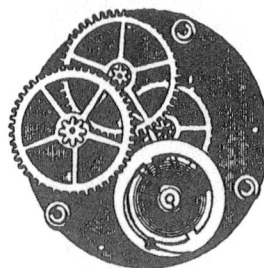

shown. Thus it will be seen that the long wind so often complained of enables the Company to make and put upon the market the Waterbury at its low price.

Why the Waterbury takes so long to wind.

The Watch has a Spring nine feet long, from which the power is applied direct. This Spring is very thin, is also very evenly tempered and

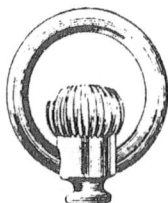

almost impossible to break. This secures a uniform and steady movement and avoids expensive compensation and regulating.

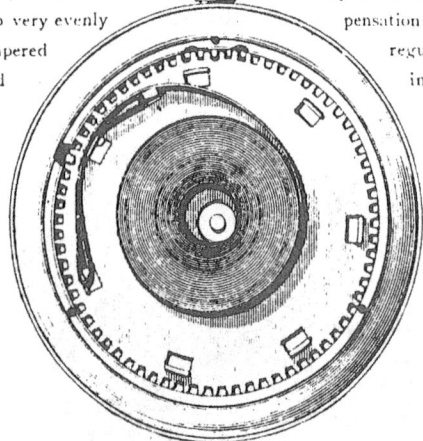

Watch with Bezel off showing Winding

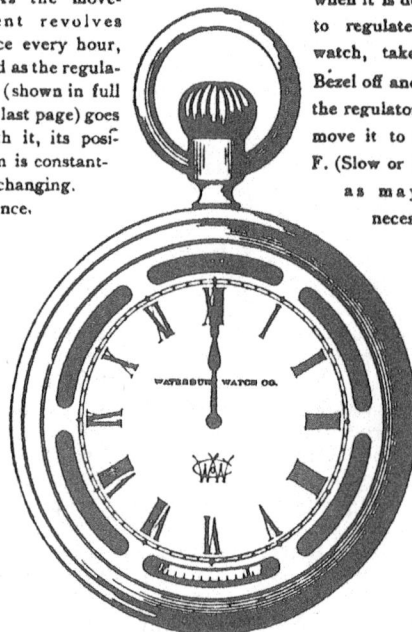

How to Regulate the Waterbury.

As the movement revolves once every hour, and as the regulator (shown in full on last page) goes with it, its position is constantly changing. Hence,

when it is desired to regulate the watch, take the Bezel off and find the regulator and move it to S. or F. (Slow or Fast) as may be necessary.

WATERBURY WATCH CO.

Pinion (at top) and Regulator (at bottom).

80. Waterbury descriptive material explaining their long-wind watch.

icut clock industry and it was only a matter of time, so to speak, before economical punched parts would reduce the cost of watches. Hopkins' Auburndale rotary had been a very near miss. Its bridgework was l a m i n a t e d from stamped parts and the construction was fairly simple. But Hopkins' design had fundamental difficulties, primarily with a large internal gear. Benedict & Burnham understood stamping, clockmaking and gearwork, so rejected Hopkins' watch. Next time around, Locke and Merritt knew all the better what to look for. Buck, perhaps at their urging, returned to a fullplate layout rather than bridgework: two simple plates. His design had only 58 parts and very simple construction. While having similarities to the Auburndale, it avoided an internally toothed gear. The escapement, usually the most complicated part of a watch, was Buck's cleverly designed duplex, with the supporting bridges slightly flexible so that the escapement could be adjusted. Finally, all parts, except steel arbors, were designed to be punched from rolled brass, like low-cost clock parts which Benedict & Burnham knew how to make. This avoided a lot of complicated tools and machining operations, yet good design allowed the stamped parts to be robust and dependable. The dial and movement were skeletonized to give an interesting appearance and allow customers to see the unique movement rotating about inside. This had a quaint look, making a fascinating show of a watch that otherwise would have been merely cheap and crude. The skeletonizing gimmick may have also been an indication that at $3.50 these were as expensive as some other watches then available. This price undoubtedly dropped to a level near $3 as Benedict & Burnham increased their manufacturing capability. Irrespective of its price, this mechanism defined the dollar watch, with several features that were utilized by all makers that followed in the dollar watch market:

1) punch press manufacturing methods;
2) no jewels or adjustments;
3) clever design with few parts;
4) an integral inexpensive case.

The Benedict & Burnham watches were produced for several years, with slight variations in the skeleton design, while effort was underway to assemble a new company. This was organized as the Waterbury Watch Co. and in contrast to the pilot project, no expense was spared in preparing a new building and machinery of the first order. When the building opened in 1881 it manufactured 600 watches daily, and production eventually went over 1000. These were the beginning of the Waterbury Longwind watches which had only 54 parts compared to 150 in a conventional watch. It became the manner of the new company to designate their models by the letters of the alphabet, thus the Benedict & Burnham watches had been series A and Waterbury followed with two similar series, obviously B and C. They were all longwind models but series B and C eliminated the skeleton dials. This would imply that the efficient new Waterbury factory brought the price below $3. Waterbury watches were cheap enough that no skeletonizing gimmick was necessary, which would only have added to cost anyway. Their reputation was established, their price was unmatched, and their sales volume skyrocketed. Series G was modeled in the early 1880's as a non-rotating movement, with a lever escapement in which the pallets were made of hard steel blocks rather than jewels. These were 18 size, ³₄-plate keywind watches and a series of ten thousand were started, but few were completed. Series E was another longwind watch, Waterbury's last such model, and the company produced these by the millions. There was a large demand for them and with Waterbury's equally large manufacturing capability they became cheap enough to use as advertising items. They could even be ordered with an

advertising legend on the case if sufficient quantities were purchased. The watch had its disadvantages, mostly in the giant mainspring which gave it its name. Since the mainspring was connected directly to the rotary movement, which turned once every hour, the mainspring had to have more than 24 coils in order to run for 24 hours. The result was a nine-foot-long spring that was not very convenient to wind, especially since the winding stem turned only in one direction. Longwind models, for which Waterbury was known as the city of everlasting spring, were finally discontinued in 1891 for a more conventional watch, still using Buck's duplex escapement. The company continued with a general line of inexpensive watches, some with a few jewels, till they were reorganized as the New England Watch Co. in 1898. At the peak of their production they made 1500 watches per day, and more than any other company, pioneered a method of watchmaking which put a useful timepiece within the grasp of every person. Waterbury had begun the dollar watch era, though the price of their watches never reached one dollar. Their use of stamped brass construction created a useful pocket watch at the lowest possible price and "cheap" watches soon became the largest segment of American watchmaking.

It is interesting to speculate that this all might have happened back in 1840. An important feature of Henry Pitkin's "American Lever Watch" was use of stamped brass plates. Pitkin's genius had leapt far beyond watches of his day, directly to the most economical means of manufacture. Combined with the simplicity of a keywind watch, his design could have been very low cost if he had lasted long enough to perfect it. It would easily have been cheaper than hand-made imports or products of contemporary pioneers of machine production, who were trying to make traditional-style movements. James and Henry Pitkin failed due to business and marketing difficulties, but their watch was a technical masterpiece. They wandered off to New York where they had few connections and great difficulty introducing their unique timepiece. Their simple stamped design was most untraditional and such watches were not readily accepted until Victorian Americans had become educated to Yankee-style watchmaking. The Pitkins manufactured the first truly American pocket watch,

81. Waterbury model G #1427 — lever escapement having pallets made of hard steel blocks rather than jewels.

THE FACTORY OF THE CHESHIRE WATCH COMPANY, CHESHIRE, CONN.

82. Factory of the Cheshire Watch Co., Cheshire, Conn., ca. 1888.

83. Cheshire movement #1660 — pendant and winder are built into the movement, requiring a special case, ca. 1883.

similar in many respects to Waterbury, or even to West-clox, the last American pocket watch.

On the basis of simplicity, the inexpensive duplex escapement continued to be made for a number of years. New England Watch Company, successors to Waterbury, made a well known skeleton duplex model as well as other styles based on old Waterbury designs. Some were among the few in modern watchmaking in which the second wheel of the train, known as the centerwheel, was not in the center of the watch, carrying the minute hand. It was likely displaced to the side because room was needed for a stemwind/set mechanism, which was relatively large for a small inexpensive watch. The other example of this unusual layout was the design used by George Roskopf for his first cheap watches. Another company using the duplex escapement was the Suffolk Watch Co. of Waltham. In 1901 they were

successors to the Columbia Watch Company, which had been formed by Edward Locke to produce an inexpensive lever watch. The Suffolk Watch Co. also made some lever watches but had several models, including the Atlas and Cambridge, which used an inexpensive duplex. This was similar to Buck's design used at Waterbury except that escape wheels were machined from a solid piece rather than stamped, which would have made them slightly more expensive. Suffolk was purchased by the United States Watch Co. of Waltham after only a brief existence, and the Knickerbocker Watch Co. of New York City was the only company other than New England to sell duplex watches in the 20th Century. They existed from 1890 to 1930, selling inexpensive watches and Swiss imports. Their American watch appeared to be a New England duplex with a custom

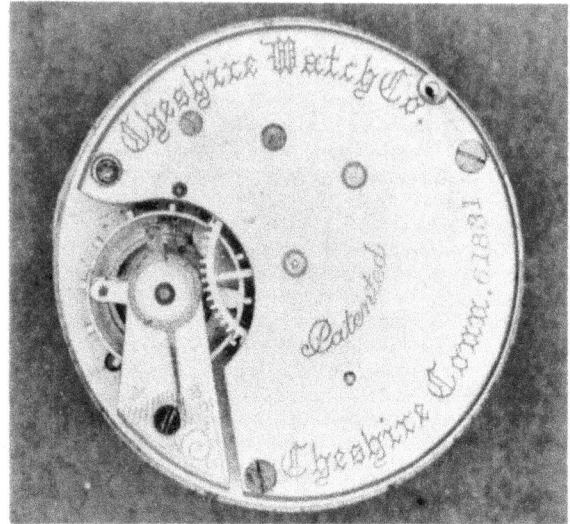

84. Cheshire movement #61,831 — designed to fit standard cases.

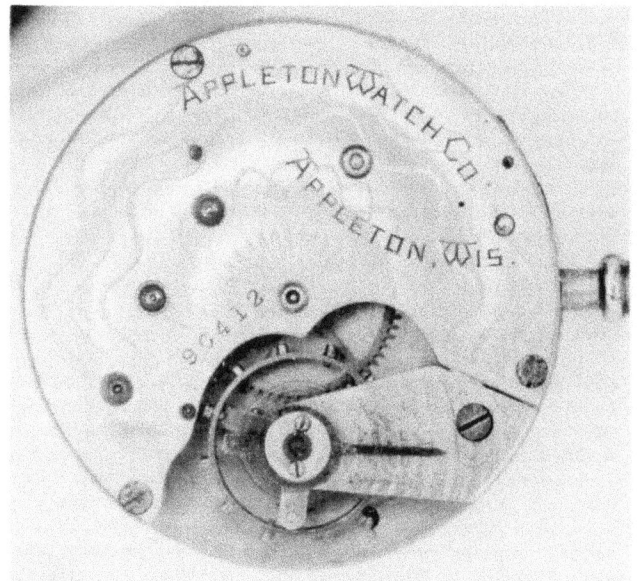

85. Appleton Watch Co. #90,412 — similar to the Cheshire watch. Made in Appleton, Wisconsin, by Remington Watch Company, using machinery from the defunct Cheshire Company.

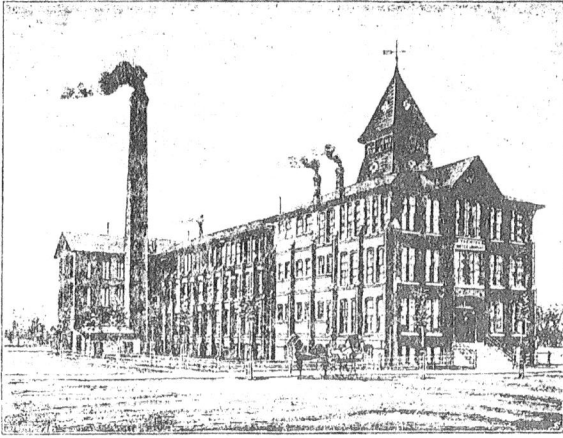

86. Trenton Watch factory ca. 1890.

design plate over the top plate of the movement. Thus they may have been a sales company rather than a true manufacturer, and actually there is little known of the company or how it operated. An interesting item with the company was that it was purchased by Ira Guilden in the early 1920's, when its business had degenerated badly, who later became vice-president of Bulova and president of Waltham. With New England, and perhaps Knickerbocker, D. A. Buck's famous escapement came to the close of a long and prosperous career. Countless millions had been made since 1878 because the design was simple to manufacture. It was on this basis that the "dollar watch" industry was launched with the Waterbury Watch Co. and grew to the mammoth proportions required to put a watch into virtually every American's pocket.

If the duplex was so cheap and easy to make, why did it fail to survive in the inexpensive watch market, for it was totally replaced by the pin pallet escapement that George Roskopf introduced? Indeed it would appear that the pin pallet was more complex and required more parts. There were probably several things which brought this about. First, since the duplex was protected by a series of patents, competitors had to find a design which they could use without lawsuit. While Roskopf may have patented his pin pallet escapement, the patents either ran out, or the American companies used it anyway and were never prosecuted. By 1900 the pin pallet had also been developed to the point where it was simple, dependable, and easy to produce. Its makers thus had no incentive to change as companies using

the duplex faded out of business, they just stayed with the pin pallet. The pin pallet, or pin lever, may also have outperformed the duplex. Inexpensive watches were usually slow-beat watches which stopped easily due to motion of the body, especially if the watch had been used for a time and was no longer running strongly. If this were to happen with a duplex watch, the watch would remain stopped since the escapement was not capable of restarting itself. However, the pin levers shared the same self-starting property as other lever escapements and such brief stoppages would never be noticed. Therefore users of pin lever watches probably got longer useful service out of their timepieces. Whatever the case may be, pin pallet watchmaking was on a comfortable footing at the turn of the century and simply continued on.

Certainly the most notable purveyors of pin lever watches were the Ingersoll brothers, Robert and Charles. In the early 1880's they had established a business selling rubber stamps in New York City, following their older brother Howard, who had invented the method for vulcanizing rubber type. They expanded their retail busines, adding departments dealing in dollar items, metal specialties, and bicycle accessories. As volume increased, a wholesale business was added and sales were amplified in 1892 by issuing a catalogue, in which they offered their first watch.[9] This was not what would be considered an Ingersoll watch, but was a Swiss import selling for $3.95. Later that year the first true Ingersoll watches appeared, wholesaling for $1 and in 1896 the Ingersoll Yankee retailed for $1. The Ingersoll brothers obtained these watches from the Waterbury Clock Company, which had begun in 1857 as a branch of the Benedict & Burnham Mfg. Company. (This becomes confusing since Benedict & Burnham also spawned the Waterbury Watch Co. in 1880.) Waterbury had previously sold such watches under their own name but enjoyed no great sales in them. Then, in 1894, Robert Ingersoll placed an order for 500,000 watches in return for becoming sole agent for any pocket timepieces Waterbury might produce. Watches became a booming business with the Ingersolls, so that they purchased the Trenton Watch Co in 1908 and the New England Watch Co. in 1914. By 1916 16,000 watches per day were being produced under the Ingersoll name in ten different models. With the recession following World War I the Ingersoll empire became financially troubled and in 1922 their watch business, including the Ingersoll name, was purchased by the Waterbury Clock Co. on the condition that neither brother would engage in the watch business again. Waterbury continued their line of Ingersoll watches till 1944, when they were reorganized as the U. S. Time Company, makers of Timex. The quantity of Ingersolls sold had been staggering, 6 million by the turn of the century,

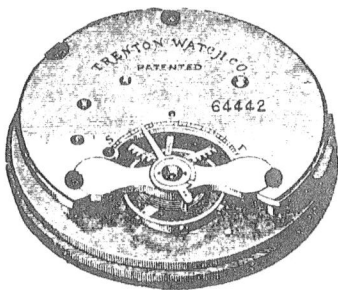

87a. Early Trenton movement based on New Haven Watch Co. design ca. 1888.

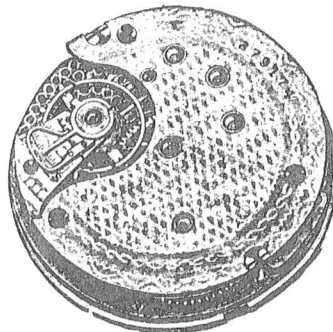

87b. Trenton ¾ plate ca. 1900.

87c. Ingersoll Trenton bridge model ca. 1910 — separate-appearing bridges are formed from one piece.

Timeline axis (years): 1876 78 80 82 84 86 88 90 92 94 96 98 1900 02 04 06 08 10 12 14

AUBURNDALE W. CO. *

BENNEDICT & BURNHAM *

WATERBURY W. CO. *

NEW HAVEN CLOCK CO. *

CHESHIRE W. CO.

NEW HAVEN W. CO.

TRENTON W. CO.

MANHATTAN W. CO.

NEW YORK STANDARD W. CO.

KNICKERBOCKER W. CO.

NEW YORK CITY W. CO. *

ROBERT H. INGERSOLL & BRO. *

INTERNATIONAL W. CO. *

COLUMBIA W. CO. → SUFFOLK W. CO.

WESTERN CLOCK CO. (WESTCLOX*)

ANSONIA CLOCK CO. *

BANNATYNE W. CO. *   E. INGRAHAM

MANISTEE W. CO.

NEW ENGLAND W. CO.

INGERSOLL *

INGERSOLL TRENTON

TO 1922
TO 1956
TO 1922
TO 1929
TO 1930
TO 1944
TO PRESENT
TO 1930
TO 1971

Fig. 18

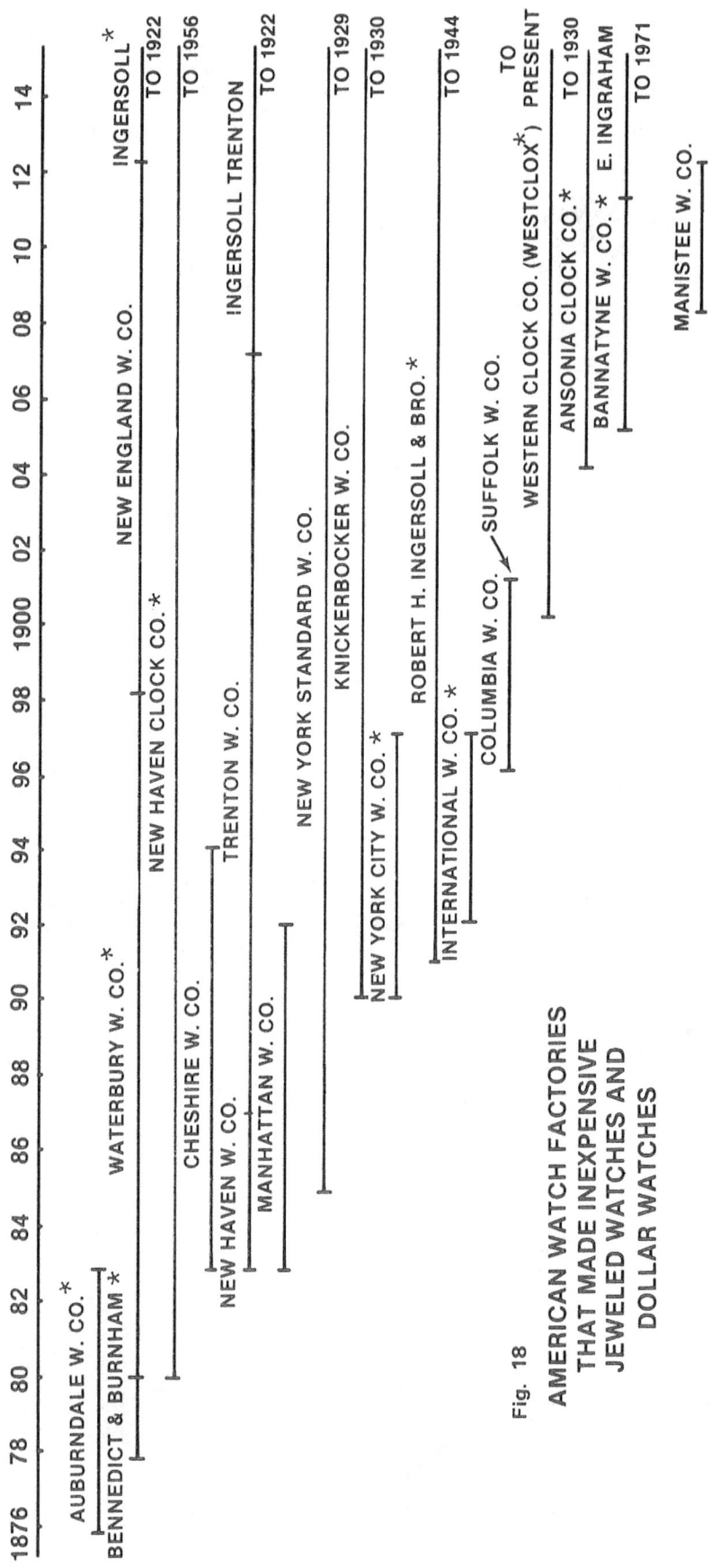

AMERICAN WATCH FACTORIES
THAT MADE INEXPENSIVE
JEWELED WATCHES AND
DOLLAR WATCHES

* DOLLAR WATCH COMPANIES

1981 © M. C. HARROLD

60

88. New York Standard fullplate ca. 1887 — identical to worm drive model except that the star-shaped cut-out has been eliminated.

89. New York Standard split plate ca. 1910 — it is nearly identical to the Keystone Howard model 11, being produced at a time when both New York Standard and Keystone Howard were o w n e d by the Keystone Watch Case Company.

60 million when they sold out in 1922, and 96 million when production ceased in 1944. It was this massive volume that built the Ingersoll reputation as "The watch that made the dollar famous."

Other companies went into the business as well, such as the International Watch Company, New Haven Clock Company, and New York City Watch Company. Like Ingersoll, several were particularly long lived. For instance in 1905 Archibald Bannatyne left the Waterbury Clock Co. to start the Bannatyne Watch Company. He was bought out in 1912 by the E. Ingraham Co. who continued making watches under the Bannatyne patent till they were sold to McGraw-Edison Co. in 1967.[10] Through the years these watches varied little from the first ones brought out by Bannatyne in 1905. Also, in 1884 several men of Waterbury went west to start the United Clock Co. in Peru, Illinois, building inexpensive alarm clocks. They went bankrupt in 1887 but their business was reorganized as the Western Clock Company, known today as Westclox, and is still in the business of making inexpensive watches.[11]

Makers of cheap watches realized the need to package their cheaper grade of machinery in a stylish case in order to attract the public. Generally in the watch trade, movements were examined and purchased separately from cases in order that the buyer could decide for himself how to divide his money between mechanism and its protective covering, the case. The buyer was aided in this by the system of standard sizes; so when the movement was chosen, there was no problem finding the desired case to fit it. In offering a watch for less than $2 there was little to be gained from this exercise, since there were no options in selecting the movement and a normal silveroid case alone may have cost a dollar. All told, the way to put the most inexpensive product on the market was to have the movement and case designed and built in the most economical manner by the same company, which often meant part of the mechanism was built into the case. For instance, the winding/setting mechanism in inexpensive watches often had some of its pieces assembled into the case, and Waterbury longwind watches combined part of the winding click in the case construction. Also the dial was frequently combined with the case. And of course, the cases had to be cheaper than standard cases; snap backs and bezels were used instead of hinged or screw backs, and they often included celluloid crystals and paper dials. All these things were different from the standard American watch case. Anyway, the movement of a cheap watch was not the most enchanting sight to entice a customer; he was more likely to be interested if his attention was drawn away from the mechanism toward a reasonably

attractive case and dial. The best bet was to assemble the most attractive package possible and offer it complete. This occasionally invited visual gimmicks. Best known was the Waterbury longwind skeleton model, with its gothic patterns punched from the dials and movements, allowing customers to see the movement rotating hourly inside. New England Watch Co. also tried a skeleton watch, but of more conventional construction.

While dollar watch companies were driving down their prices, other companies were developing inexpensive jeweled watches in the $5 to $8 range, making a slightly more durable, accurate timepiece available for only a bit more money than a dollar watch, then selling for $3.50. (There can be a lot of confusion over prices. Those used here are retail, which the individual customer pays to their local jeweler. Old trade catalogues usually list wholesale prices, at which retail jewelers buy from wholesaling houses, and which are roughly half of retail price. Watch factories sold only to wholesalers, and until this century, wholesalers sold only to retailers. Wholesaling facilitates distribution of mass-produced goods; the large retail markup on jewelry pays the interest on money invested in inventory which may sit on the shelf for years, since jewelry is a slow turnover business.) In 1883 were formed the Cheshire Watch Co. in Cheshire, Connecticut, the New Haven Watch Co. in New Haven, Connecticut, and the Manhattan Watch Co. in New York City. The Cheshire company attracted Mr. D. A. A. Buck as superintendent and marketed an 18 size, ¾-plate, gilt movement which they sold cased for $8.[12] Since the movement would fit only their own cases they also came out with a movement which would fit standard American cases and thus be more attractive to the wholesale trade. The New Haven Watch Co. removed to Trenton, New Jersey, before getting into production and became the Trenton Watch Company. They manufactured a line of inexpensive jeweled watches which they sold in cases, like the Cheshire watch, for about $7.50,[13] and also made some chronograph watches. They were absorbed by the Ingersoll brothers in

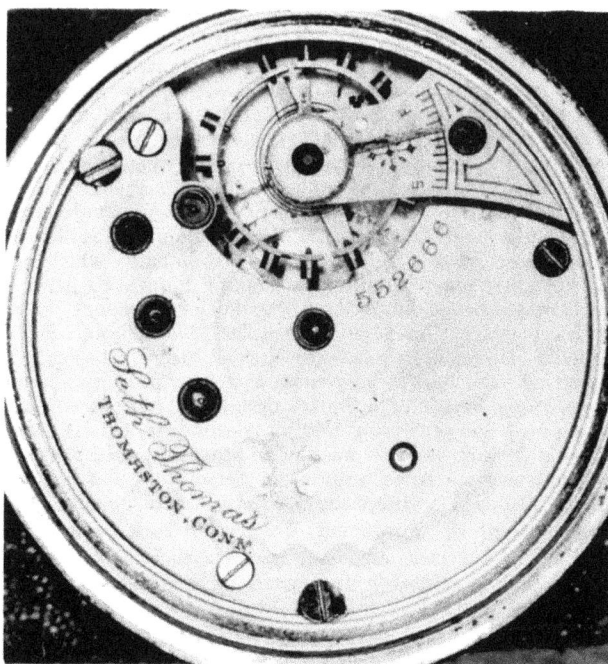

90. Seth Thomas ¾ plate #552,666 ca. 1900 — inexpensive 7-jewel movement.

1908, becoming Ingersoll Trenton Watch Co. and continuing with a similar product as before. The Manhattan Watch Co. also sold cased watches of a low-grade variety in 18 size, both hunting and open face, some of the latter being sweep seconds and with a start/stop mechanism.[14] The Manhattan watches used an interesting escapement with upright steel pallets and were unusual in their casing, setting and stop actions. Following these companies, the New York Standard Watch Co. was formed in 1885 and came out with an inexpensive 18 size watch at about $7. This featured the unique worm drive escapement, intended mostly as a gimmick to sell watches rather than keep time, which was discontinued for a conventional lever escapement after a brief period. The firm persisted a bit longer with their unusual winding/setting mechanism, utilizing a large internally toothed gear the diameter of the entire movement, placed between the pillar plate and dial.[15] Seth Thomas, which began watchmaking in 1884, did a major part of its business in inexpensive jeweled watches. This product was based primarily on their long lever ¾-plate movement. Over the years, they avoided the image of being inexpensive watchmakers by promoting their line of conventional jeweled watches. Through mass-production by such companies as these, the price of inexpensive jeweled watches was down to $5 by 1900.

While such companies were making solely inexpensive watches, large diversified companies such as Elgin and Waltham got the price of their plain movements below $7. Because of their manufacturing capability, large factories could offer low-cost watches of a slightly better class than others in the field. Their sophisticated machinery produced a mechanism with better finish on the working parts, where better finish made a better watch, as well as better adjusting, temperature compensation, and quality of winding/setting mechanism. Combined with an inexpensive nickel case, a person could purchase a quality pocket timepiece for $7, for which he had a finer watch than the erratic hand-made articles the Swiss had been selling at the same price.

There appeared near the end of the century a very interesting watch, which was almost American, employing a number of inexpensive watch features. It was the invention of Albert H. Potter, a somewhat restless individual but the finest American watch artisan there was. During the 1850's and 1860's he resided in New York, Cuba, and Chicago, then in 1872 went to Geneva, Switzerland, where he remained till his death. Though he is chiefly remembered for hand-made watches of exquisite quality, he sold patent rights to the New Haven Watch Co. for an unusual watch intended to be of the best quality possible under $4. These rights reverted to the Trenton Watch Company, which apparently had every intention of making the watch, but for some reason never did. Around 1894 Potter had about 15,000 of these watches made in Charmilles, Switzerland, and a number of them did come to the United States. The project must not have been a huge success, for nothing more ever came of the design. The Potter design had very few parts and quite a clever winding/setting mechanism, but the most notable feature was the manner in which the backplate of the movement curved around to form the sides of the case.[16] To this watchplate/case piece were secured the frontplate of the movement, dial, bezel, back cover, and pendant of the case, resulting in a watch of fewer parts than usual, considering the movement and case together. The Swiss watches were very well built, with high quality and excellent finish throughout, mostly in 7 jewels but sometimes more. It is difficult to speculate why the Charmilles watch did not proceed to a longer career, and it might have if produced in an American factory. Quality

may have suffered a bit, but American factory methods would have made a satisfactory piece at a better price, and the watch would have been marketed more thoroughly.

The world had been waiting for dollar watches, since most of the world could not afford a watch till dollar watches became available. Waterbury longwinds were the archetype of this breed, even if their price never made it down to $1. They were forerunners and the most spectacular in their approach to dollar watch principles of simple design punched from rolled brass. This was The American Watch made The American Way and all dollar watches that followed approached these principles in some manner. George Roskopf attempted a proletarian watch through his unique design but was limited in his success by Swiss handcraft methods. The key to reducing cost, in watches or any other product, was development of cheaper manufacturing techniques. Clever design played an important role, but its prime aspect was designing around the cheapest method of production. Roskopf's ingenuity was soon overshadowed by the same force that terminated hand clockcraft, replaced wooden-works clocks, and spread Connecticut clocks to all continents: stamped brass construction. Dollar watchmaking was thus closely connected to the clock industry through such companies as Waterbury Clock, E. Ingraham, Ansonia, New Haven Clock and Westclox. Since clockmaking was strongly centered around Connecticut, dollar watches were primarily a Connecticut industry, with virtually every dollar watchmaker being there except Westclox. Even Westclox could trace its origins back to Connecticut, though it was established in Illinois, which produced more conventional jeweled watches than any other state.

Dollar watches, combined with the rise of inexpensive jeweled watches, generated a burst of activity in watchmaking. Between 1879 and 1890 the number of United States watch companies doubled from 10 to 20, mostly through growth of the low-priced industry. The "Gay 90's" were troubled times financially, eliminating some of these newly formed concerns, but the first decade of this century saw the number of dollar watch companies rise again. The seven watchmaking survivors in 1932 included four dollar watch companies, New Haven Clock, Waterbury Clock, Westclox, and E. Ingraham. It seems odd now that all this excitement concerned watches that still cost a day's wages or more to buy, but that just shows the luxury level of watches in those times. The difficulty in getting prices that low had been monumental and competition among dollar watchmakers was keen. A low price meant a low profit margin and business stability was based on high volume. Consequently these companies were stepping on each other's toes seeking new business, and "cheap watches" could be a nerve wracking industry.

This created pressure on conventional jeweled watch companies also. There had already been some competition among the established firms such as Waltham, Hampden, Elgin, Illinois, and Rockford, plus the Swiss. Along with recessions during the 1870's this had caused reductions in prices and development of cheaper grade movements. The advent of dollar watches brought demand for yet lower prices, and inexpensive jeweled watches were in direct competition with the conventional jeweled watch firms. Therefore these latter companies knuckled down to a serious effort, exemplified by a major overhaul of Waltham management and practices in 1883 aimed at higher production of lower-cost movements. This aggressiveness from the industry "heavies" made life difficult for inexpensive jeweled watchmakers, few of which survived and none of which were significant producers. Only Trenton and New York Standard had careers of any magnitude, along with Seth Thomas, which had sales of conventional watches as well.

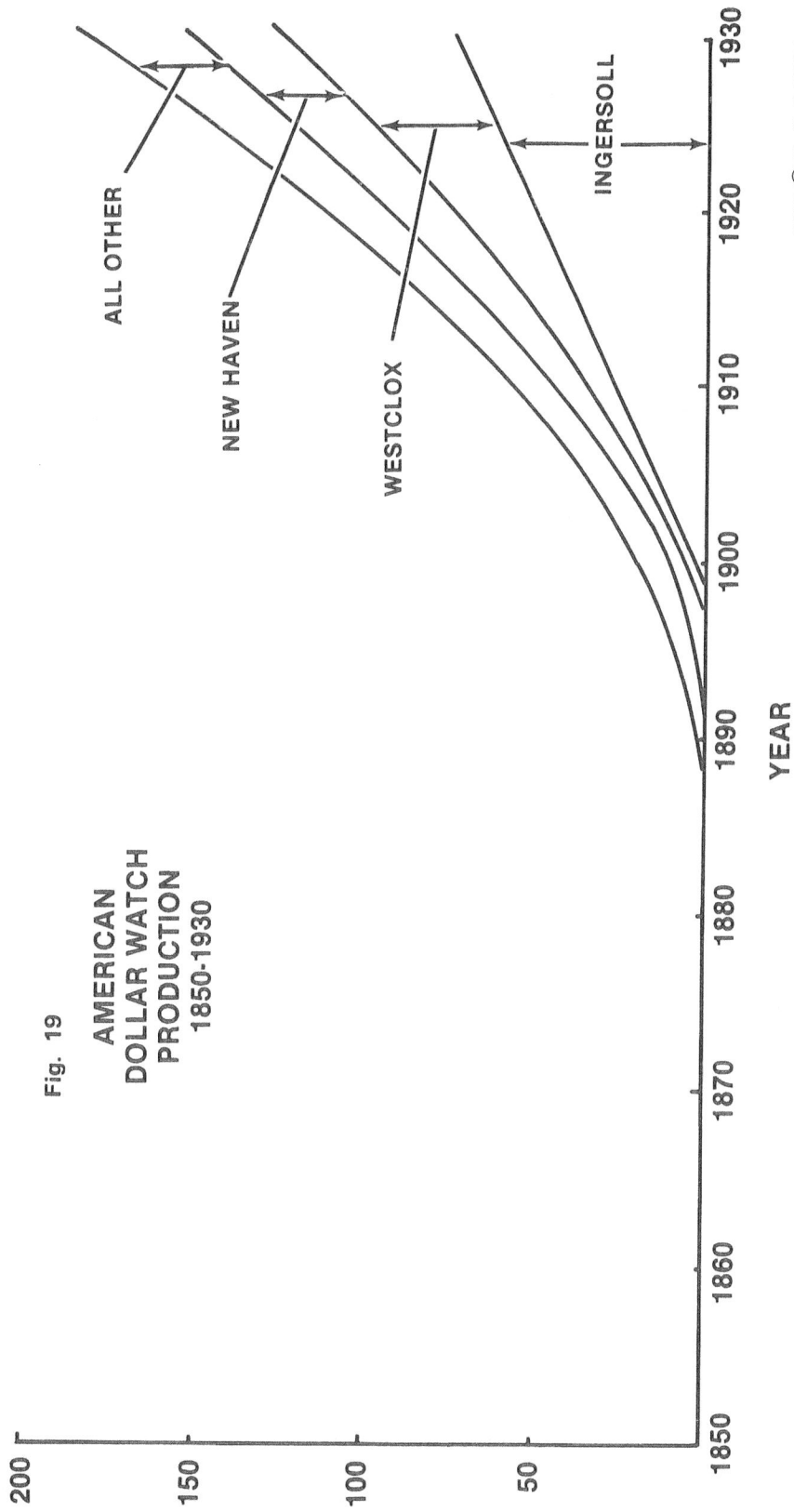

Fig. 19

AMERICAN
DOLLAR WATCH
PRODUCTION
1850-1930

ALL OTHER

NEW HAVEN

WESTCLOX

INGERSOLL

1981 © M. C. HARROLD

200

150

100

50

WATCH PRODUCTION · MILLIONS TO DATE

1850   1860   1870   1880   1890   1900   1910   1920   1930

YEAR

63

The result of this effort was millions of watches. Conventional jeweled watch companies such as Elgin manufactured large quantities, but dollar watch production rates were much higher, from the very beginning, and the quantity of such watches was immense. The comparison below makes this clear.

Total Watch Production — In Millions — To Date

Year	1880	1890	1900	1910	1920	1930
Jeweled Watches	3	11	23	41	63	80
Dollar Watches	1	5	18	58	118	186

It was as though watches had been invented all over again. Watches first appeared in 1500, being spring-driven clocks small enough to carry. In 1880 stamped brass clockmakers began producing cheap spring-driven clocks small enough to carry and begot a whole new industry. Clock companies and dollar watches took over American watchmaking. After 1900 half the companies and most watches belonged to this league. Auburndale almost made it happen, Waterbury led the advance, and Ingersoll made the dollar famous.

REFERENCES

1. Willis I. Milham, *Time & Timekeepers* (New York, The Mac Millan Co., 1941), p. 396.

2. Charles S. Crossman, *The Complete History of Watchmaking In America* (reprinted Adams Brown Co., Exeter, NH), p. 30.

3. Jaquet & Chapuis, *The Technique and History of the Swiss Watch* (Olten, Switzerland, Urs Graf-Verlag, 1953), p. 187.

4. Paul M. Chamberlain, *It's About Time* (New York, Richard R. Smith, 1941), p. 72.

5. Crossman, p. 169.

6. Chris Bailey, *Two Hundred Years of American Clocks and Watches* (Englewood Cliffs, NJ, Prentice Hall, 1975), p. 210.

7. Crossman, p. 148.

8. Ibid., p. 152.

9. Robert F. Tschudy, *Ingersoll "The Watch That Made The Dollar Famous"* (BULLETIN, National Association of Watch and Clock Collectors, April 1952), p. 97.

10. George Townsend, *The Watch That Made The Dollar Famous* (Alma, Michigan, George Townsend, 1974), pp. 8-11.

11. Ibid., p. 28.

12. Crossman, p. 165.

13. Ibid., p. 167.

14. Ibid., p. 186.

15. Chamberlain, p. 104.

16. Ibid., p. 449.

VI

PRECISION WATCHES

While the young American watch industry was striving to produce economical timepieces during the 1860's, it also aimed to cultivate excellence. High-grade watches were needed to raise the reputation of factory work, making American watches indisputably desirable rather than just cheap alternatives. For watches, excellence obviously meant accurate timekeeping, which implied numerous bits of precision mechanism behind the dial. This, in turn, required that companies develop machines to manufacture higher quality parts and introduce fresh capability to manufacture parts that had previously been imported. In addition, since the American system of standardized sizes had resulted in movements being retailed separately from cases, factories had to develop an aesthetic style for watch mechanisms, which were thoroughly examined by the customer. So, in the years just before and after 1870, watch factories were emphasizing capability to mass-produce:

    a) precision balances and hairsprings;
    b) accurate wheels and pinions;
    c) nickel damaskeened finish;
    d) sure-acting stemwind mechanisms.

Most importantly, these all required advancements in machine capability. The industry thereby added quality to its growing reputation, became independent of foreign balances, hairsprings and jewels, and acquired machine techniques which could then be utilized in all watch grades. This last point had a large impact on the trade, for once properly developed, a machine could economically mass-produce parts of excellent value. With such capability, the industry could manufacture watches of increasingly dependable and uniform quality in all price ranges. This drive for precision was therefore indispensable to the American factory system, in competition with hit-or-miss handwork methods used on low-cost imports.

The earliest advocates of precision had been E. Howard and the Nashua Watch Company. Howard experimented extensively to develop a precision timepiece he could produce and machinery to produce it, ultimately relying on hand finishing to complete his watches. The Nashua venture was dedicated to machine-made excellence. It was perhaps fortunate that Nashua went broke, to be absorbed by Royal E. Robbins, America's consummate watchmaking entrepreneur, into his growing giant at Waltham. Anticipating high-grade competition from Nashua, Robbins' company had begun around 1861 to produce its KW18 model in top-quality 15- and 19-jewel grades named Appleton Tracy & Co. and American Watch Co. ("American" is completely spelled out on this grade). After purchase of Nashua, the KW18 was used solely for lesser-quality watches. Nashua's ¾-plate movement became Waltham's KW16 and KW20, and the Nashua department at Waltham manufactured these in the American Watch Co. grade, using its superior machinery to develop some of the finest American watches. Robbins marketed these in a manner to elevate his company's reputation, and that of all American watchmaking.

As additional companies entered the industry they faced noteworthy competition at Howard and Waltham. When Elgin movements appeared in 1867, they were B. W. Raymond models of fine quality, accompanied by a retail price of nearly $50, that gained Elgin a sound footing in the railroad market. United States Watch Co. of Marion, New Jersey, likewise marketed watches of high standards. While these and other new companies began business with gilded fullplate movement designs which then dominated American watchmaking, the trend in high-grade watches shifted another direction. Howard utilized a split-plate pattern and the Nashua design at Waltham was based on an English ¾-plate style. Growing prestige, and price tags, of Howard and American Watch Company grade movements created a distinct class of American gentlemen's pocket watches, superior to and different from fullplate workhorse timekeepers for the rail trade. These gentlemen's timepieces were intended for the prestige market occupied by such imported watches as Charles Frodsham and Jules Jurgensen, where a plain timekeeper sold for approximately $250. Following the Nashua model, ¾-plate styling became standard for American premium watches. United States Watch Co. introduced a ¾ plate patterned after Swiss watches by Charles Jacot and August Saltzman.

91. Waltham KW18 #36,157 Appleton Tracy & Co. grade high-quality model made by American Watch Co. ca. 1861 to compete with Nashua and Howard.

92. Howard #306,723 high-grade American stem winder ca. 1880. Howard production had settled into a ¾-plate design ca. 1860.

93. Waltham #501,532 American Watch Co. grade 19-jewel model 68 ca. 1871.

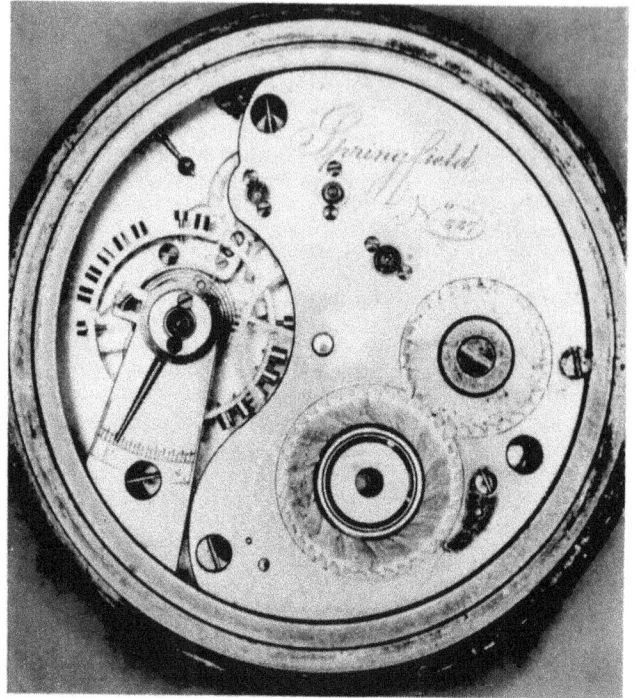

95. New York Watch Co. #337 Springfield grade 19-jewel stem winder ca. 1871.

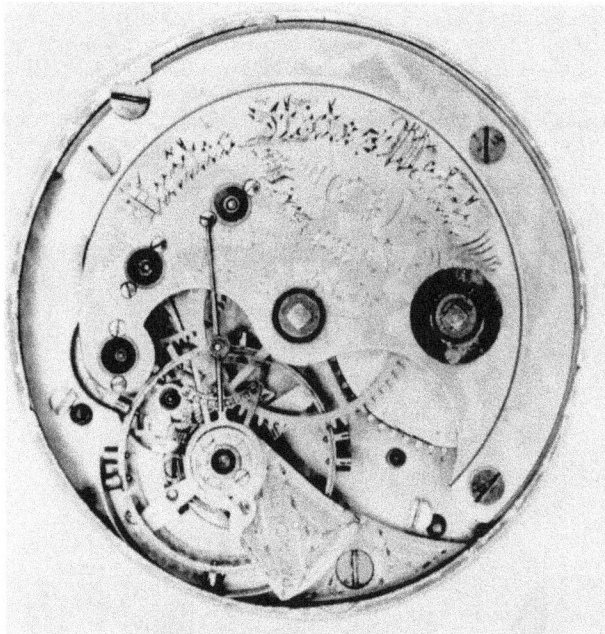

94. U.S. Marion #53,528 United States Watch Co. grade 15-jewel, ca. 1873, was retailed at over $300. A 19-jewel version of the United States Watch Co. grade was marketed for $400.

96. Adams & Perry #2042 21-jewel ca. 1876 — material has been marked Lancaster Watch Co. who were successors to Adams & Perry.

97. Waltham #501,503 — American Watch Co. grade 19-jewel model 68 ca. 1871.

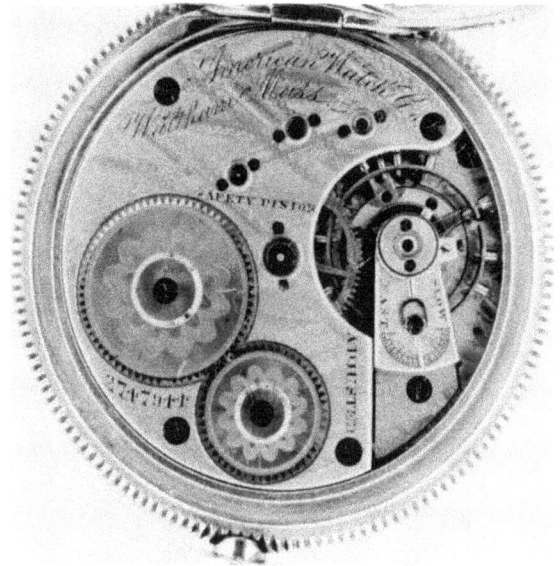

98. Waltham #2,747,944 — American Watch Co. grade 21-jewel model 72 ca. 1885.

(Their 19-jewel United States Watch Co. grade, with unique vernier regulator, was made in both ¾ and fullplate movements that were marketed for $400.) New York Watch Co. chose a gilded movement style similar to that of German watches by Lange and Grossmann. Adams and Perry used a ¾-plate design of their own. (Elgin refrained from entering the prestige market till the late 1870's, eventually introducing their premium grade convertible model.) There was a preference during this period for using 19 jewels in premium timekeepers, being the standard 17 jewels plus cap jewels on the escape wheel. The chief exceptions

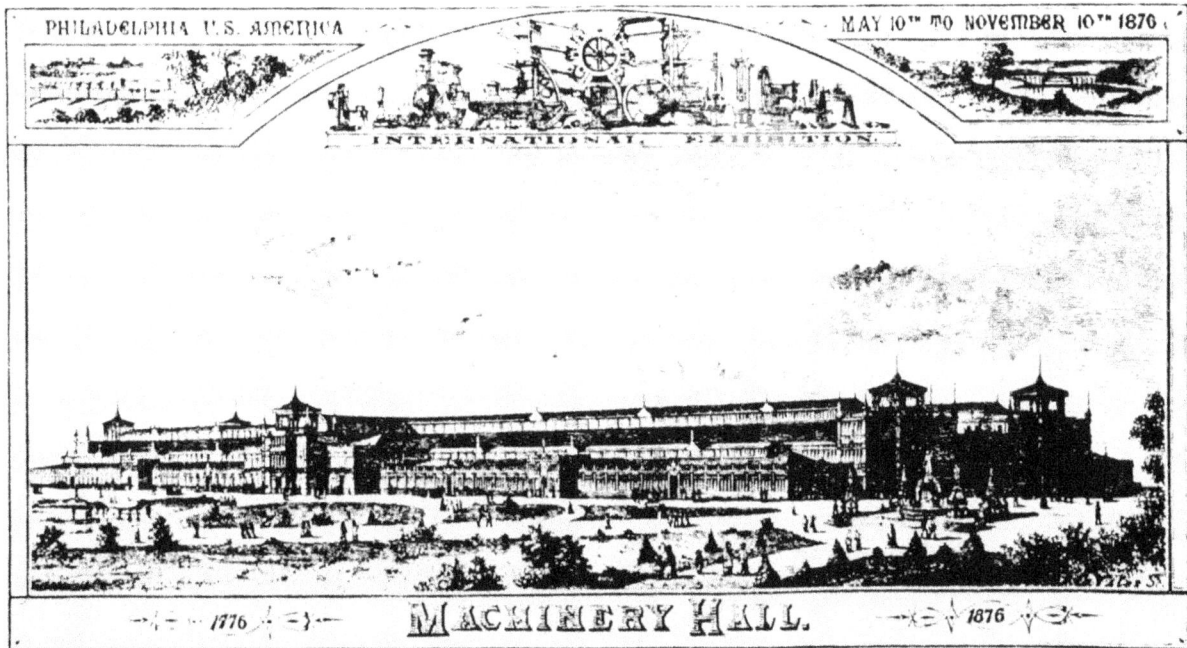

99. Machinery Hall of the Philadelphia Centennial Exhibition.

100. Waltham #670,039 — American Watch Co. grade 19-jewel model 72 ca. 1872.

101. Waltham #5,000,110 — American Watch Co. grade 19-jewel model 88 ca. 1891.

to this were that the bottom jewel was sometimes eliminated from the center wheel, following European practice, and Howard normally used 15 jewels. As these movements began appearing, during the late 1860's and early 1870's, nickel finish and damaskeening were adapted from Swiss watch styles, damaskeening being mechanized into intricate patterns to suit gingerbread Victorian taste. Gold wheelwork and jewel settings were also borrowed from the Swiss, and modern gadgets such as stemwinding and safety pinions were added. By the early 1870's, premium-grade American watches were elaborately designed showpieces, having high-grade balances, escapements, and hairsprings to make them noteworthy timekeepers.

There was additional emphasis to produce prestigeous watches in this period since a world display was being planned for the American Centennial in 1876. The Great Exposition of London in 1851 had introduced the "World Fair" concept, to be followed by similar English and French fairs in the 1850's and 1860's. These were primarily industrial exhibits to promote products of the host nations, though foreign entrees were also invited. The Philadelphia Centennial Exhibition was therefore organized to demonstrate American agriculture, invention, and industry as emerging world influences. Visitors saw massive American steam engines, Edison's telegraph, sewing machines, typewriters, and a curious audio device shown by Alexander G. Bell. They also saw automatic machine tools used increasingly in the American system of interchangeable part manufacture. Some of the most fascinating of these were exhibited by the American Watch Company, turning out finely-finished screws, pinions, and staffs in completely automated fashion. The capability of such machinery was reinforced by premium-grade watches exhibited by several companies, with displays tauting American quality and factory methods. Perhaps the most spectacular of these were 19- and 21-jewel model 72's featured by the American Watch Company. (At the Sydney Exposition in 1889, one of the 19-jewel 72's was prize timekeeper against such notable competition as Victor Kullberg, Nicole Nielsen, Adolph Lange, and Louis Audemars.)

The impact of these exhibits was perfectly clear to Edward Favre Parret, Swiss member and Chief Commissioner of the International Jury on Watches at Philadelphia. He was certain that automated watchmaking would overrun handwork methods, so that the Swiss must either mechanize or perish. He therefore accumulated sample American materials for illustrating speeches designed to motivate the Swiss trade. As he reported to his countrymen:

"Gentlemen, here is what I have seen. I asked from the manager of the Waltham Company a watch of a certain quality. He opened before me a big chest. I picked out a watch at random and fixed it to my chain. The manager asked me to leave the watch with them for three or four days that they might regulate it. 'On the contrary,' I said to him, 'I want to keep it just as it is to get an exact idea of your workmanship.'

"On arriving in Locle (Switzerland), I showed the watch to one of our first adjusters . . . who took it apart. At the end of several days he came to me and said literally, 'I am astonished, the result is incredible. You do not find a watch to compare with that in 50,000 of our make.' This watch, I repeat to you, gentlemen, I myself took offhand from a large number, as I have said. One can understand by this example how it is that an American watch should be preferred to a Swiss watch."[1]

These were not hollow words, for by 1880 Swiss sales in America had fallen sharply and American dollar watches were rapidly displacing inexpensive Swiss cylinder watches. Since Swiss economy centered around export, Swiss watchmakers accepted the burden of reorganizing their methods in order to survive.

So, by 1880 the drive for precision had achieved its desired results; homemade watches dominated the vast American marketplace. Half of them were stemwind since mechanisms could easily be machine fabricated. Added to price cuts, American factories were offering more for less as competition increased and the Swiss were eliminated. Nickel damaskeening, temperature compensation, and adjusting were becoming available in all but the cheapest grades. American horology's reputation, established at the Philadelphia Centennial, built a solid demand for its products. Advertisements expounded that even Swiss watchmakers admitted the superior quality of American workmanship.

The Centennial period had been a golden age for American watchmaking. Precision and quality had been their own ends, resulting in exquisite new watches and elaborate new machines. With competition gripping the industry, it was the machines that carried on, spreading new-found elements of excellence across the trade. Having proven itself, the industry felt less compelled to manufacture showpiece watches after 1876, leaving little American competition for Howard, 21-jewel Elgin convertibles, and the American Watch Co. grade Waltham, the last of which was reduced to the Riverside Maximus grade by 1900. Needless to say, none of the premium timepieces had ever been made in great quantity, for there was severely limited demand for $250 watches in 1880. Such quality was good for company reputations, but it had to be diluted to larger quantity and lower price to build company profits. While thousands of average watches began enjoying faint reflections of premium timekeepers, the railroad market offered substantial sales potential for quality watches.

Table VII
Retail Prices ca. 1885

Type of Watch	Price	Description
American Watch Co. grade	$250	Model 72, 21J, 18K case
Elgin grade 91	$250	21J convertible, 18K case
Railroad grade 15J	$40	silver case
Common 15J	$20	silver case
Common 7J	$10	nickel case
Inexpensive 7J	$6	Cheshire, N.Y. Std.
Waterbury	$2.50	dollar watch

The Northwest Territory, now the area from Ohio through Wisconsin, had been attracting settlers during the 1820's and 30's and was beginning to produce natural resources and agricultural excesses. This resulted in a need for transportation, both locally and back to the east coast. Indeed, the United States was expanding all over the continent and had a dire need of transportation routes to join its far flung states and territories. Not surprisingly, the first efforts in this regard were to link major water routes, joining the eastern seaboard to the Hudson River and beyond to Lake Erie. In the western states the Great Lakes were the focal point, with routes radiating down to the Ohio River and over to the Mississippi. Early in the century, canals were seen as the most promising means of transportation; boats were easily capable of carrying large loads and could be readily towed along by the most universal source of energy, animal power. One of the early triumphs of canal building was completion of the Erie Canal, which connected the Hudson River to Lake Erie in 1825. The trend continued through the 1830's and 40's during which midwestern states built canals up and down their length, such as two in Ohio from the Ohio River up to Lake Erie and a giant waterway the entire length of Illinois with a branch over to Chicago. The latter was the Illinois and Michigan Canal and when completed in 1848 Chicago became the only port on the Great Lakes with a through waterway to the Mississippi, which meant it was a major junction between the Mississippi and the east coast. The subsequent rise of Chicago as a trade center was automatic.

The 1820's and 30's also saw interesting developments in England. The industrial revolution was breaking into transportation with railroads for hauling freight, using huffing steam engines as the means of locomotion. With England's small size only 300 miles of track could tie together major industrial towns between Liverpool on the west coast and London on the English Channel. By contrast, America had rugged terrain and huge expanses to cross, so that many people considered rail transportation impractical here, especially with primitive engines available in the early years. But railroad advocates were not easily dismayed and the 1830's were a decade of bold expansion. Rails crossed South Carolina to the Savanah River, Massachusetts to the Hudson, and a railroad headed west out of Baltimore to the Ohio River. Technical improvements as well as aggressive track layers pushed across the Appalachians and the west began to fill with ties and iron. In 1848, when the Illinois and Michigan Canal opened, the Galina and Chicago Union Railroad became the first railway into Chicago, heading west across the top of Illinois through the little town of

Elgin. In 1851 rails joined the Atlantic seaboard to the Great Lakes. By 1852 there was a through connection of sorts on to Chicago, and by 1860 11 railroads centered around the Chicago area. The American Civil War was the first war in which fighting waged around major rail centers, and with the war effort Chicago began to develop as an important manufacturing center. Industry joined agriculture to boost the already massive transportation system in the area; Chicago became second only to Pittsburgh in steel production, and the first steel rails in the country were made there in 1865, replacing iron. All this promoted the expansion of railroads further west, having eliminated canals as a means of large-scale transportation, and in 1869 the continent was spanned by track with the golden spike driven at Promontory Point, Utah.

With the growth of railroads it became apparent how vast America was, and one of the aspects by which this stood out was time. Due to our geography, railroads ran primarily east and west, which of course involved a time shift. Since there was no established system of standard time in the country, railroads usually based their schedules on local time at the company headquarters, or some major city served by the line. But local time meant time as kept by a sundial, and since the sun took a long time to travel across a country this large, there were significant discontinuities between various schedules. There was a shift of 20 minutes across the state of Pennsylvania, 45 minutes from New York to Chicago, and more than three hours from Boston to San Francisco. One company's 10:00 arrival could chug into town after another line's 10:10 departure had already left, which made for some difficulty doing business with the railroad network.

There was more to the railroads' concern with time than

TABLE
SHOWING THE TIME AT MORE THAN ONE HUNDRED DIFFERENT PLACES
When it is 12 o'clock, Mean Noon, at Boston, Mass.
Compiled from the most reliable data.

Place	H. M. S.	Place	H. M. S.	Place	H. M. S.
Albany,	N. Y. 11 49 15	Hartford,	Conn. 11 53 31	Paris,	France. 4 53 35
Alexandria,	Egypt. 6 45 46	Havana,	Cuba 11 11 43	Philadelphia,	Pa. 11 43 34
Annapolis,	Md. 11 58 45	Haverhill,	Mass. 11 59 54	Pike's Peak.	9 44 14
Astoria,	Oregon. 8 29 26	Havre,	France. 4 44 39	Pittsburgh,	Pa. 11 24 06
Augusta,	Me. 12 01 54	Honolulu,	S. Is. 6 12 46	Portland,	Me. 12 03 16
Baltimore,	Md. 11 57 47	Jeddo,	Japan. 2 04 14	Portsmouth, N. H.	12 01 11
Bangor,	Me. 12 09 06	Key West,	Fla. 11 17 04	Providence,	R. I. 11 58 39
Berlin,	Prus. 5 37 49	Leavenworth, Kan.	10 25 18	Provincetown,	Ms. 12 04 03
Boston,	Mass. 12 00 00	Lexington,	Ky. 11 07 02	Quebec,	C. E. 11 59 25
Brattleboro',	Vt. 11 54 06	Lima,	Peru. 11 35 41	Queenstown,	Ire. 4 11 14
Bremen,	Ger. 5 19 26	Liverpool,	Eng. 4 49 13	Raleigh,	N. C. 11 29 02
Brooklyn,	N. Y. 11 49 20	Lockport,	N. Y. 11 29 10	Richmond,	Va. 11 34 21
Buffalo,	N. Y. 11 28 34	London,	Eng. 4 43 50	Rio Janeiro, Brazil.	1 51 38
Canton,	China. 12 17 20	Louisville,	Ky. 11 02 14	Rochester,	N. Y. 11 32 50
Cambridge,	Mass. 11 59 44	Lowell,	Mass. 11 58 58	Rome,	Italy 6 34 08
Cape Town,	5 58 10	Melbourne, N.S.W.	2 24 07	Sacramento,	Cal. 8 38 23
Charleston,	S. C. 11 24 32	Memphis,	Tenn. 10 43 42	Salem,	Mass. 12 00 40
Chicago,	Ill. 11 12 43	Mexico City,	Mex. 10 07 02	Salt Lake City, Ut.	9 15 50
Cincinnati,	Ohio. 11 06 16	Milwaukie,	Wis. 10 52 37	San Francisco,Cal.	8 34 27
Cleveland,	Ohio. 11 16 59	Mobile,	Ala. 10 52 08	Saratoga,	N. Y. 11 49 14
Clinton,	N. Y. 11 42 37	Montreal,	Vt. 11 53 50	Savannah,	Ga. 11 19 53
Columbus,	Ohio 11 12 02	Montreal,	C. E. 11 59 02	Southampton,Eng.	4 38 38
Concord,	N. H. 11 58 18	Nantucket,	Mass. 12 03 52	Springfield,	Mass. 11 54 13
Constantinople.	6 40 30	Nashua,	N. H. 11 58 28	St. Louis,	Mo. 10 43 13
Detroit,	11 12 04	Newark,	N. J. 11 47 34	St. Petersburg.	6 45 28
Dover,	N. H. 12 00 38	N. Bedford,	Mass. 12 00 32	Stockholm, Swed.	5 56 29
Dublin,	Ireland. 4 18 52	Newburyp't,	Mass. 12 00 46	Syracuse,	N. Y. 11 39 26
Eastport,	Me. 12 16 30	New Haven,	Conn. 11 52 32	Tallahasse,	Fla. 11 05 50
Edinburgh,	Scot. 4 31 30	N. London,	Conn. 11 55 54	Taunton,	Mass. 11 59 50
Fall River,	Mass. 11 59 46	New Orleans,	La. 10 44 14	Toronto,	C. W. 11 26 41
Frankfort,	Ky. 11 05 31	Newport,	R. I. 11 59 10	Trenton,	N. J. 11 45 08
Geneva,	Switz. 5 08 50	New York,	N. Y. 11 48 14	Utica,	N. Y. 11 43 22
Gibraltar,	Spain. 4 22 50	Niagara Falls, N.Y.	11 27 58	Vienna,	Aust. 5 49 46
Gloucester,	Mass. 12 01 35	Norfolk,	Va. 11 39 00	Washington,	D. C. 11 36 02
Greenwich,	Eng. 4 44 14	Northampton,	Ms. 11 53 44	West Point,	N. Y. 11 48 24
Halifax,	N. S. 12 29 47	Omaha City,	Ne. 10 20 18	Wheeling,	Va. 11 21 26
Hallowell,	Me. 12 04 54	Oswego,	N. Y. 11 37 50	Woodstock,	Vt. 11 54 10
Harrisburgh,	Pa. 11 35 54	Panama,	C. A. 11 25 17	Worcester,	Mass. 11 57 01

BOSTON:

PRINTED BY ANDREW HOLLAND.
Wood Cut and Oil Color Printer,
20 Washington Street.

KILBURN AND MALLORY, ENGRAVERS,

102. Time in more than 100 places based on noon-time at Boston. From a Waltham advertising pamphlet ca. 1862. Most cities listed are in the United States, making obvious the need for standard time.

just arranging deliveries and connections. Most legs of the rail system had only a single line of track to handle a number of trains each day, going in both directions. Occasional sidings left room for trains to pull off the line, allowing oncoming trains by or faster trains to pass from behind. But some means was required to inform trainmen when to pull onto sidings. Various schemes were tried but the most common was simply to organize a schedule dictating where trains would be at all times and when passings were to occur. To assure safety, schedules had to be very strictly obeyed and trainmen needed to carry accurate watches synchronized with the company master clock. Some system was therefore necessary to coordinate all company clocks and watches. One such method was "The Adams System of Time Records" created by John C. Adams to assure that watches on the line were accurate and on time. It was apparently being used by some companies in the 1880's but does not seem to have been widespread or long lasting.[2] The problem of schedules was further complicated by passage of trains across other companies' tracks. Coordinating safety schedules was just too difficult when each company had its own time.

Superintendents of the various lines finally met in St. Louis in 1872 to arrange summer schedules. Out of this simple beginning grew the Time Table Convention, which became the American Railroad Association, and finally The Association of American Railroads, which convened the General Time Convention at Chicago's Grand Pacific Hotel in 1883. At that meeting they threw out a host of local railroad times and adopted one system of standard time zones for Canadian and American use. There were five zones, each centered about meridians at 15-degree intervals from Greenwich, England, so they were even hours apart from Greenwich Mean Time. In 1918 the U.S. Congress passed the Standard Time Act by which the railroad time zones became law, and daylight saving time was adopted to conserve electricity. Thus those first railroad time zones are roughly the time zones in general use today.[3]

It can be seen there was to be a close relationship between watches and railroads. In fact, based on the rise of

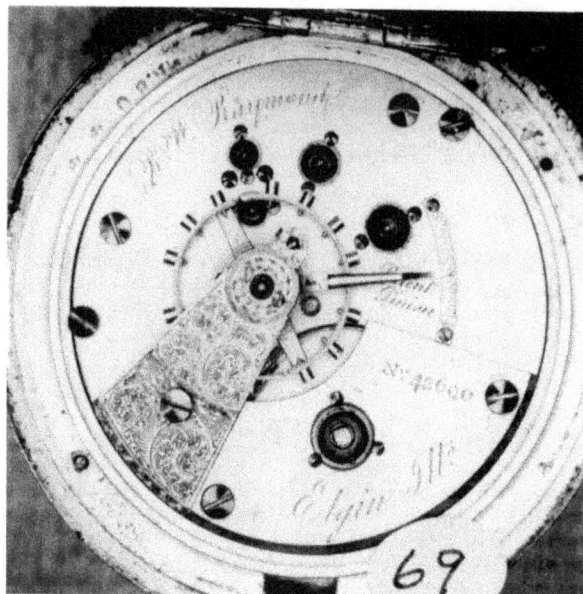

104. Elgin B. W. Raymond #42,600 ca. 1869 — 15-jewel railroad grade.

railroads and industry around Chicago, J. C. Adams organized the National Watch Co. in Elgin, Illinois, in 1864 and went on to form three other watch companies in the state: Cornell, Illinois, and Peoria.[4] For the same reasons, Illinois was to have numerous other companies, or attempted companies: Rock Island, Rockford, Freeport, Western, Aurora, and Westclox. So while watchmaking was considered a New England industry, Illinois became a major watch producing area. It produced more conventional jeweled watches than any other state and had a good share of the inexpensive watch market. Except for the latter, much of the business was built around the railroad trade.

English and Swiss watches were of course being used for rail service as the American watch industry began in 1850. English pieces were generally considered more capable timekeepers. Those movements were in the traditional gilded fullplate pattern, and it was not uncommon for dials to be marked "Railroad Timekeeper." (In later years, such a designation almost certainly indicated that a watch was not of railroad quality.) There were few, if any, formal requirements for such watches to meet, but qualified specimens probably did not exceed one minute error per week. They would have been temperature compensated, adjusted for position errors, and adjusted for isochronism, though some fusee watches may not have had the last. Jewel counts would have ranged from 9 to 21. Quality watches were commonly 9 to 15 jewels during that period, though some watches for the American market utilized cap jewels on escapements, elevating numbers to 17 or more

As Howard and Waltham, the latter fortified by purchase of Nashua, began marketing quality timepieces around 1860, American watches likely began appearing in railroad service. One clearcut milestone was 1866, when the Pennsylvania Rail Road ordered model 57 movements from the American Watch Co. of Waltham, in the Appleton, Tracy & Co. grade with 15 jewels.[5] When the first Elgin movements appeared the following year, they were the railroad grade B. W. Raymond, similar in appearance to Waltham's model 57. These were both gilded fullplate movements in the style that trainmen respected, and as more American watch companies were formed, other like movements en-

103. Dial of English watch by E. D. Johnson marked Railroad Timekeeper. It is a high-grade movement, perfectly capable of railroad timekeeping performance, ca. 1850. In later years, such a designation was used to decorate low-quality watches.

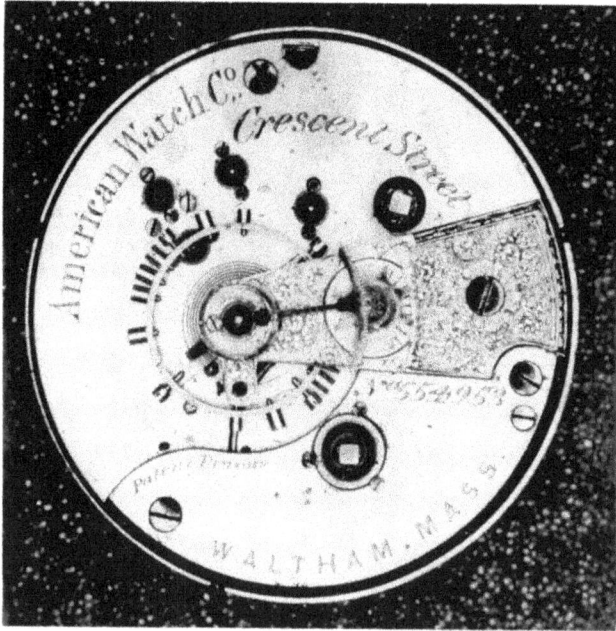

105. Waltham model 70 #554,953 ca. 1870 — 15-jewel movement designed for the railroad market, with micrometer regulator and both key wind and set from the back, which is uncommon on a fullplate movement.

106. Illinois Bunn #240 ca. 1872 — 15-jewel.

tered the rail trade. Through the 1880's, railroad timekeepers thus remained the upper tier of common style watch mechanisms, appearing nearly identical to average $15 movements, but with refinements that brought prices to about $40. As with their foreign predecessors, American railroad watches were temperature compensated and adjusted for position and isochronism. Flat hairsprings were nearly universal through the 1860's and 70's since Breguet springs were not yet mass-producible and Swiss springs were not heat treated, leaving them unstable over an extended period. Mainspring barrels in railroad grade movements frequently utilized stopworks to reduce isochronal error over the winding period, and 15 jewels became the accepted standard. So while railroad models in these years were not spectacular, they were serviceable mechanisms, designed to be reliable workhorses of moderate cost.

In 1870, Waltham introduced the model 70 Crescent Street as their highest grade fullplate movement, designed specifically for the rail trade and among the earliest watches with a micrometer regulator. (Howard was using Reed's patented whiplash regulator.) A few of these also carried a push-set stemwind mechanism, but most watches remained keywind during the 70's. Quick trains became increasingly common, which resulted in more rapid oscillations of the balance wheel. Early watches had been slow beat, oscillating 4½ beats per second. Quick train movements vibrated 5 beats per second, which produced more stable timekeeping. (Precision wristwatches today oscillate at 6 beats per second.) By 1880, stemwind and damaskeened nickel finish were growing popular. Railroad movements were still 15 jewels and fullplate, but nickel finish, Breguet hairsprings, stemwind, leverset, quick trains, and safety pinions were becoming expected, while the price remained around $40. Gold fill was replacing silver as the popular case material in the 80's, and jewel settings were frequently of gold.

As the 1880's wore on, railroad timekeepers were finer and more handsome than common watches, but not markedly different. They were a quality nickel finish fullplate, typified by the Waltham model 83. In that period, a rail grade timepiece was likely to have come from Elgin, Waltham, Hampden, Illinois, Rockford, Peoria, or Columbus. Howard did some business in the rail trade but less than other companies since his prices were markedly higher. The Hampden Railway and Perry, Illinois Bunn, Elgin B. W. Raymond, and other railroad grade movements had all been improved over the years, keeping with 15 jewels, but growing more accurate and sophisticated. Such watches were used religiously by trainmen, for which railway timekeepers took on a glamor that they have never lost.

Despite the glamor, there was a disastrous wreck in 1891 on the Lake Shore and Michigan Southern Railroad, apparently due to the malfunction of a watch. A fast moving mail train collided with a slower train near Kipton, Ohio, killing both engineers and nine mail clerks and touching off an investigation.[6] Concerned railroad officials asked Webb C. Ball, a Cleveland jeweler, to organize a standard system of checking timepieces. Ball found that while a number of railroad personnel carried respectable timepieces, others used all manner of poor watches, dollar watches, and even alarm clocks, since there were no regulations mandating minimum requirements.[7] For railroads to assure that adequate timekeepers were in service, they clearly had to establish standards and enforce them. In that way they would not have to rely on the discretion of individuals to be sure that good watches were in use, in proper regulation, and repaired when needed. Ball set up a system of standards for the Lake Shore and Michigan Southern and eventually became General Time Inspector for a number of railroads, with offices in Cleveland, Chicago, San Francisco, and Winnipeg. The Ball firm was paid by railroads to oversee the standards, while Ball, in turn, supervised local watch inspectors along the rail lines. These inspectors were not paid by Ball at all but were jewelers who received their compensation in revenue for watch sales and repair work, and being a Ball inspector could considerably enhance business. Many railroads established their own inspection departments rather than contract with an outside individual like Webb Ball, although they generally followed Ball's standards and methods.[8]

71

107. Rockford #20 ca. 1875 — 19-jewel and micrometer regulator.

109. Waltham model 83 #3,908,585 ca. 1890 — 15-jewel railroad grade. The model 83 was one of Waltham's most widely produced pocket watches.

108. Rockford #5258 ca. 1875 — 15-jewel railroad grade.

110. Hampden Railway #619,741 ca. 1890 — 15-jewel.

111. Waltham model 83 #6,018,101 ca. 1895 — 17 jewels,

113. Waltham model 83 #7,452,120 ca. 1898 — 17 jewels.

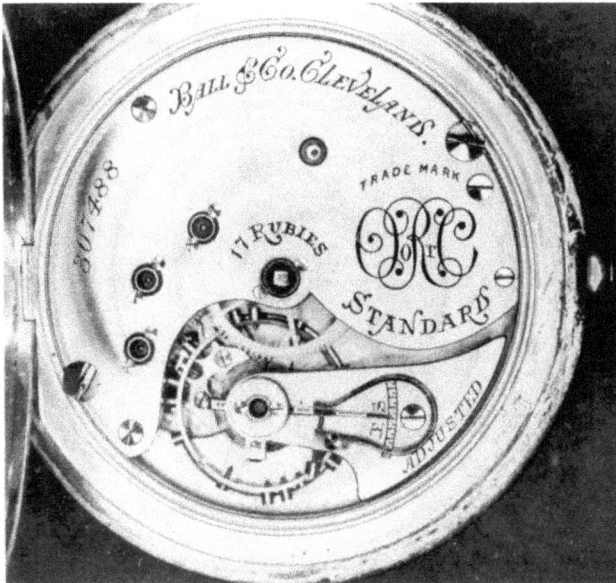

112. Howard #307,488 made for Ball ca. 1895 — ¾-plate watches were unusual in the railroad market.

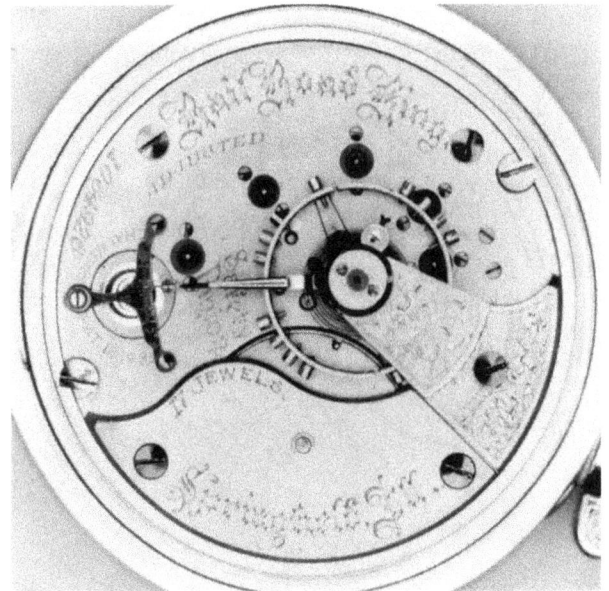

114. Illinois Rail Road King #1,064,926 ca. 1900 — 17 jewels.

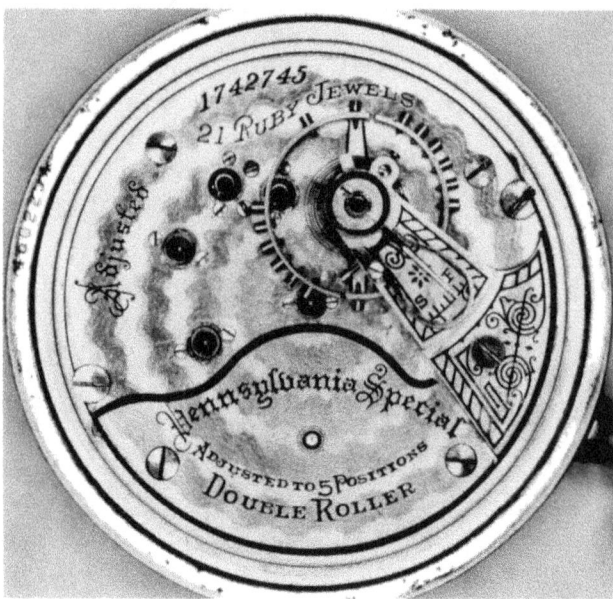

115. Illinois Pennsylvania Special #1,742,745 ca. 1903 — 21 jewels.

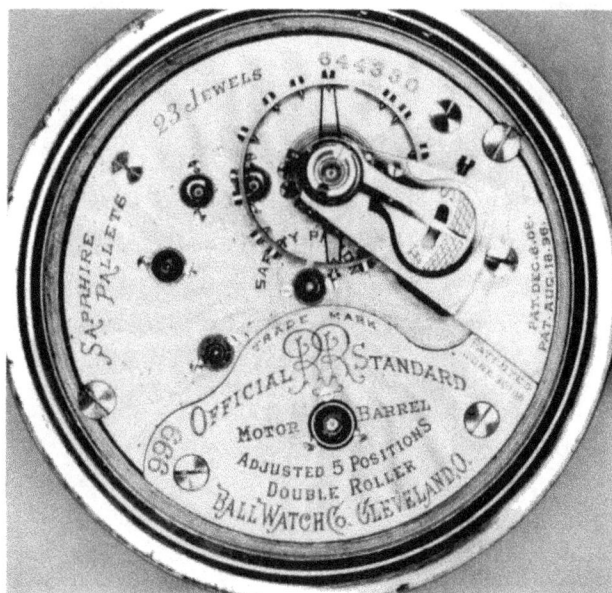

117. Hamilton #644,330 ca. 1907 — made for Ball, 23 jewels.

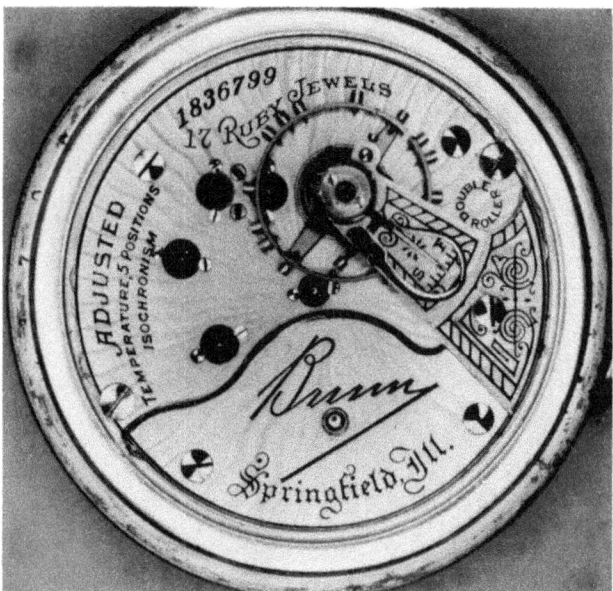

116. Illinois Bunn #1,836,799 ca. 1905 — 17 jewels.

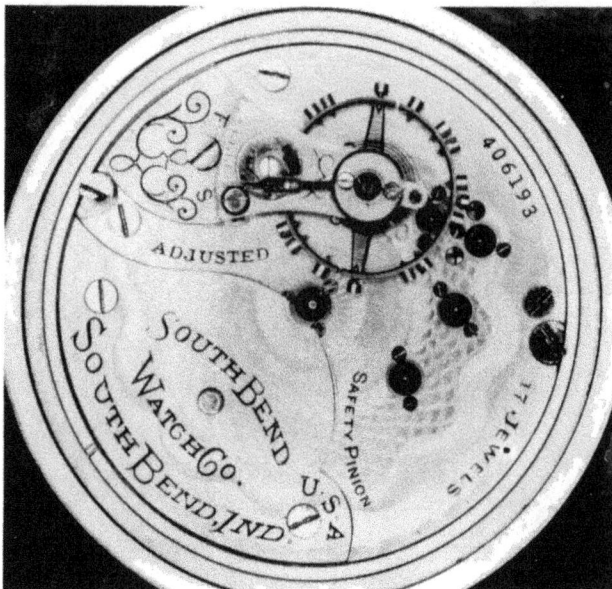

118. South Bend #406,193 ca. 1912 — 17 jewels.

Variations aside, the standards generally required that railroad watches were as follows:[9]

1. American-made 16- or 18-size movements;
2. 17 jewels minimum;
3. open face, with the stem at 12 o'clock to preclude a hunting movement in an open case;
4. lever escapement with double roller and steel escape wheel;
5. equipped with overcoil hairspring and micrometer regulator;
6. adjusted for temperature, isochronism, at least 5 positions;
7. lever setting;
8. accurate to 30 seconds per week, and reset to correct time whenever the error exceeded 30 seconds;
9. cleaned annually and inspected every two weeks, with performance noted on a card carried by the trainman;
10. mounted in a dust-tight case;
11. equipped with a crystal free of chips and scratches;
12. equipped with a dial of Arabic numerals, heavy style hands, a second hand, and minute divisions.

The intent of all this was to provide quality timekeepers that were clear and easy to read, familiar to jewelers who would work on them, and easily repaired. The standards specified companies and models so that trainmen knew what was allowed and jewelers knew what parts to stock, which was probably the reason for requiring American watches. The English and Swiss were certainly capable of making watches that met the timekeeping requirements. However, each watch was individually crafted, and there would have been difficulty getting quality repair parts custom made in a cattle town along the rail route. Railroads realized they would be depending upon a large fleet of watches and would need replacement parts everywhere in the country, that average repairmen could use with good results. Foreign watchmakers could not fill this requirement with individually-made watches, no matter how fine their quality. The practical problem of maintaining thousands of watches in hard use would have eliminated foreign competition if the standards had not.

The standards spawned a new generation of watches. (Existing watches which met the timekeeping requirements were allowed to remain in service since sufficient quantities of new models could never have been produced fast enough for sweeping introduction.) Companies seeking the railroad business updated older models or came out with new ones. In addition the Hamilton Watch Co. was formed in 1892 with the idea of making watches to the new requirements, and was followed by South Bend and Keystone Howard. Also, Webb C. Ball had watches made to his own specifications by several companies, marked with his trademark, and sold under the name of the Ball Watch Co. of Cleveland, Ohio.

Railroad watches were required to be identified with the maker's name, for which purpose Ball was considered a maker. He registered his trademark, patented the plate designs, and put specific requirements on watches made for him. Apparently Ball watches had larger balances and weaker mainsprings than the regular watches manufactured by his suppliers. This is somewhat in keeping with Swiss practices and may have been symptomatic of Ball's appreciation for fine Swiss timekeepers. Ball 18-size watches were mainly by Hamilton, though some were pro-

119. Typical railroad watch dial — bold upright Arabic numerals and heavy hands.

120. Dial of Ball rail watch — exceptionally fine hands, note that all three hands match.

121. Hamilton 992 #790,181 ca. 1907, 21 jewels — among the most popular split plate railroad watches.

122. 25-jewel Seth Thomas Maiden Lane model #351,050 ca. 1897 — one of the watches in the post-standard jewel packing competition. Maiden Lane was one of the principle streets in the New York jewelry district, where most major watch companies maintained offices.

duced by Elgin, mostly in 17, 19, and 21 jewels. The 16-size models were Hamilton and Waltham 17-, 19-, 21-, and 23-jewel movements with some 23-jewel Illinois.[10] Ball watches were not marked with the manufacturer's name and care is required in identifying their source, especially with Illinois models. His railroad watches were most often marked "Official RR Standard" but the firm also marketed non-railroad grade watches, including 12 size, which were marked "Commercial Standard."

The number of jewels required to make a precision watch was always a subject of debate, with 17 the usually prescribed minimum and 23 the maximum. Most common was probably 21, the number quoted by Webb Ball as the useful limit.[11] It will be recalled that the standard 17-jewel watch had 7 escapement jewels plus jeweled bearings on the lever, escape wheel, third, fourth and center wheels. In going up to 19 jewels it was most common to put cap jewels on the escape wheel, but some models jeweled the mainspring barrel instead. The 21-jewel watches were almost universally the standard 17 plus cap jewels on the lever and escape wheel. Finally, 23-jewel watches were the typical 21 with a jeweled barrel. Of course there were variations on this recipe. One Keystone Howard model had jeweled banking pins, which hardly seem necessary. In an effort to attract customers, companies such as Illinois, Rockford, Columbus, and Seth Thomas found ways to incorporate more jewels during the decade after issuance of the standards. Illinois, followed closely by Rockford, introduced a 24-jewel fullplate model, which was countered by Columbus and Seth Thomas with 25-jewel fullplate versions. Illinois then marketed a 26-jewel watch, after which Seth Thomas responded with a movement encrusted in 28 jewels. That culminated what had been primarily an advertising campaign, for few of these watches had been produced, and 21-jewel, 16-size railroad watches became accepted practice.[12] Americans had always liked jewels in their watches, and through the 19th Century imported watches often had higher jewel counts than their makers would have used for the European market. English watches seldom used more than 15 jewels while marine chronometers, the apex of horology, were traditionally 11 jewels. Be that as it may, American watches of the period commonly had 15 jewels and intro-

duction of the railroad standards served to boost the average even higher. In fact the standards were probably a major step in promoting the idea that quality of watches was strictly a function of jewel count, and the competitive jewel stuffing that took place among companies in their higher-grade models fueled this notion. Many of the 19-jewel models were among the finest timekeepers in railroad service, and additional cap jewels only added to the effort of cleaning watches since they all had to be removed in the process.[13] The Hampden Watch Co. is credited with the first 23-jewel watches and this was the reasonable limit. The mystique surrounding watch jewels concerned the fact that they were gemstones, which caused confusion over their value, especially since watches were sold through the jewelry trade where gems were also obtained. In reality, watch jewels were an industrial use of gemstones and functional jewels in a watch had little dollar value.

Regulations concerning dials and hands, as well as stipulations that the stem must be at 12 o'clock and cracked crystals be replaced, were all to avoid confusion in reading correct time. Likewise lever setting guaranteed that the hands would not be accidentally moved while in the pocket or when pushing in the stem, as sometimes happens on a stemset watch. Hands were universally a heavy spade style after the standards. Dials always had upright Arabic numerals for the hours and had minute divisions, usually with some identifying feature for the five minutes. One odd variation which never became too popular was the Ferguson dial patented by L. B. Ferguson of Monroe, Louisiana. This had large five-minute numerals around the outside and smaller numerals for the hours around an inside circle, somewhat the reverse of the usual dial. This was frowned upon because it was basically abnormal and could have caused confusion.[14] Much more common was the Montgomery dial patented by Henry S. Montgomery, first general

123. A particularly bold railroad watch dial by Hamilton.

124. Hamilton model 950 #1,022,190 ca. 1908, 23 jewels.

"MINUTES AT A GLANCE."

FERGUSON RAILROAD DIAL

Mr. Jobber

WRITE TO US

Mr. Jeweler

Approved by   Railroad Officials.

This is a profitable proposition for both the Jeweler and the Jobber. The Dial retails for $2.50, and we have provided a liberal discount for the trade. We have a complete assortment on hand and will fill orders the very day received. Be sure to specify this dial on new watches, the factories are prepared to furnish it and there is sure to be an immediate demand, being thoroughly practical and a great convenience. All of the figures are vertical, perfectly plain and distinct, the minutes are largest because they are the most important.

FERGUSON DIAL COMPANY (Inc.)   -   -   -   MONROE, LOUISIANA, U. S. A.

125. Advertisement for the Ferguson dial — it can be seen that large 5-minute numerals dominate the hour numerals.

77

126. Typical Montgomery dial on a South Bend watch — upright minute figures around the chapter ring and careful inclusion of the hour numeral at 6, in the seconds bit. The 6 was often eliminated by dial makers.

128. Combination of up-down indicator and 24-hour dial.

127. Variation of Montgomery dial, which closely resembles the 1920 patented design by Montgomery.

129. Unusual railroad dial by Waltham — minutes marked by widely separated numerals.

130. Illinois split plate model made for Ball, to Ball's design. Waltham's version of this model is virtually indistinguishable from the Illinois, and Illinois used a similar design of their own (see illustration #133).

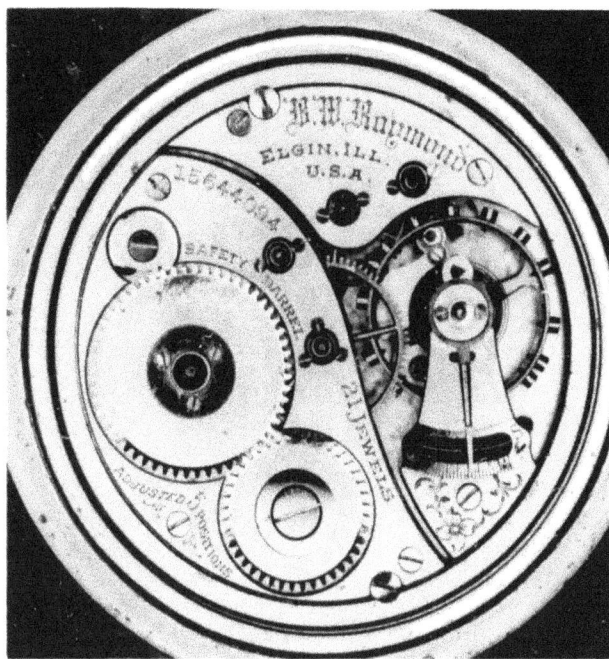

131. Elgin B. W. Raymond #15,644,094 ca. 1910, 21 jewels.

time inspector for the Atchison, Topeka, and Santa Fe Railway, which had its own inspection system. The Montgomery dial had the usual distinct hour figures but also had each minute division numbered with upright figures, and slightly larger numerals for the five minutes. He claimed that this was superior for trainmen, who were more interested in the nearest minute than the nearest hour, and that with his dial they could easily read the nearest minute.[15] Various patent dials were made up with the names of watch manufacturers marked on them, and trainmen could specify these as new or replacement parts for their watches. Twenty-four-hour dials, with 13 through 24 around an inside track of hour numerals, were not allowed on U.S. railroads but could be found on Canadian lines.

Another feature that manifested itself on the dial was the winding indicator, sometimes called an up/down indicator. This was a small sector in which a hand indicated how many hours the watch had run since its last winding. Navigational chronometers commonly used winding indicators so that navigators could check that the chronometers were not running down, leaving them stranded at sea without a time standard. Self-wind wristwatches occasionally used winding indicators also, so that the owner might know if stemwinding had become necessary. (In this instance it was really a winding reserve indicator.) Winding indicators appeared on railroad watches around 1900, when jewel stuffing competition was rampant, and were probably intended mostly as a sales feature. They entailed extra gearing, adding $5 to $10 to the cost of a watch, a substantial sum in that day. Consequently, winding indicators were never in large production.

Fullplate 18-size movements were still favored when the standards were issued in 1892, but 16-size split-plate models gained steadily in popularity and completely replaced fullplate watches by 1920. In fact, introduction of railroad standards generally coincided with the rise of split-plate style watch movements in American factory production. From its beginning, the American industry had mirrored English fullplate and ¾-plate designs, since their reputa-

tion had stood above Swiss bridge model watches. At the same time, English watchmaking was withering, and its past glory was of little concern to 20th Century Americans. By 1900, people wanted a good American watch, which was whatever trainmen said it was. Quite likely, railroaders led the advance of split-plate model watches, prompted by retail jewelers. Railroad standards required frequent cleaning, on top of which, watches in hard service would need repair. This prospect would have generated a desire in jewelers and repairmen to have bridge style movements, which were easier to work with than full or ¾-plate models. Rather than follow the Swiss practice of supporting each wheel under an individual bridge, American factories combined most of the train under a single plate, separate from the barrel bridge. In many models this plate was contoured to appear as several bridges. Extra plate work was expensive to manufacture, which was reflected in higher prices for split-plate models than for comparable f u l l p l a t e watches. Both factories and consumers would have resisted higher prices, unless they were being promoted by middlemen such as retail jewelers. The jewelry trade was very powerful in steering watch trends, and through them, split-plate model watches were one of the probable side effects of railroad watch standards.

The rise of railroad watches did create a new level of quality in the general watch market by placing a definition on what constituted a good timekeeper. This brought about more awareness on the part of the public and shifted demand to higher grade pieces, thus making them available to everybody at reasonable prices, not just trainmen. This also created a market in fakes, low-grade watches marked "Railway Timekeeper," "23 jewels," bearing the picture of a locomotive, or some other gimmick since there have always been unrealistic people to be lured by a low price. But in general railroad watches demonstrated the capability of the American watchmaking system. Were it not for the sophisticated industry, introduction of the standards would have been unrealistic, for handcraft artisans could

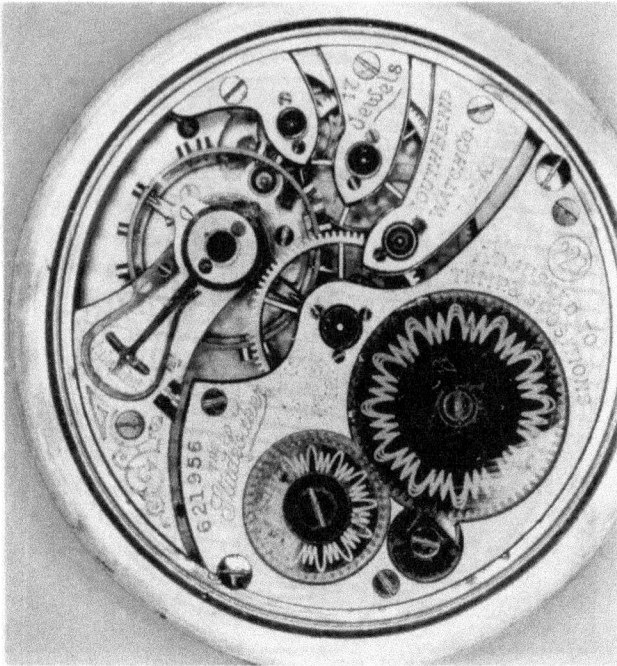

132. South Bend #621,956 ca. 1915, 17 jewels.

134. Latter-day Waltham Vanguard 23-jewel ca. 1950 — it can be seen that the sparkle is gone from one of the last survivors of American watchmaking.

133. Illinois Bunn Special #5,568,465 ca. 1925 — 60-hour winding period, i.e. a 2-day watch.

135. Waltham model 99 #9,503,677 — last American Watch grade model ca. 1900, 21 jewels.

136. Waltham model 99 #12,656,307 — Riverside Maximus grade ca. 1905, 23 jewels.

138. Waltham Premier Maximus #17,057,292 ca. 1910 — high quality with austere finish compared to the Riverside Maximus.

137. Premier Maximus dial: up-down indicator, graceful numerals, finely proportioned hands.

139. Keystone Howard high-grade 23-jewel — scarce model No. 0. Two lower train bridges are formed from one piece.

not have built the large quantity of watches required in a reasonable time or at an affordable price. When railroads guaranteed to support a market for high grade watches, the industry was able to produce a large supply in just a few years, along with a nationwide network of spare parts, all at reasonable cost.

Beginning in 1908, American watch companies cycled into another phase of premium precision pocket watches, similar to its Centennial period timepieces. The Centennial period of precision had served to develop new manufacturing capability which facilitated improvements to common watch grades. This 20th Century phase was more of a defensive measure, which nonetheless produced the finest of American watches. Having stagnated since 1900, jeweled watch sales began falling after 1910, so that fortunes were declining for most companies in that business. Moreover, increasing wristwatch sales were threatening to require expensive retooling across the industry. This combination of sagging revenues and rising expenses justified a new dramatization of American pocket watches in an effort to stimulate sales.

First of these presentation watches was Waltham's Premier Maximus in 1908 (see Table VIII).[16] Their reason for introducing it was to increase sales by raising the company image. Waltham had a history of innovating such unique and well-made products as chronographs, repeaters, and their American Watch Co. grade. Perhaps because of poor marketing, these were fading from public memory by 1908, company growth was at a standstill, and profits were falling. Waltham Watch Co. had been in a somewhat confused state since the death of Royal E. Robbins in 1902. Their production organization, accounting practices, and marketing methods were out of date, as was their top management. (As a result, the Robbins family liquidated its Waltham holdings in 1909.) The organization needed sweeping managerial house cleaning plus a modern advertising campaign to rebuild its public image. Instead, they utilized the old Centennial strategy of presenting a spectacular new watch that would lend prestige to its manufacturer as well as its owner. The Premier Maximus was meant to establish Waltham as America's premier watchmaker.

Waltham already had an extra quality watch on the market. At a retail price of $180 (with 18K case), the Riverside Maximus was more than comparable to the best 23-jewel watch from Keystone Howard, and twice the price of top-quality railroad grade watches such as the Elgin Veritas, Waltham Vanguard, or Hamilton 946. At an introduction price of $250 the Premier Maximus was substantially above the second most expensive watches of its day, it was billed as the finest watch in America, and was a superior, though

140. Keystone Howard Edward Howard model #195 — frost gilt finish with blue sapphire jewels and free-sprung balance (no regulator).

Table VIII
Presentation Watches

Date	Maker/Model	Price Range	Approximate Quantity	Jeweling	Features
1908	Waltham — Premier Maximus	$250-$750	1000 estimates 600-1400	23 jewels being 6 diamonds, 17 rubies & sapphires	winding indicator optional Class A Kew rating
1912	Keystone Howard — Edward Howard	$350	300	23 sapphires	frost gilt plates, sapphire jeweling, free-sprung balance
1922	Elgin — C. H. Hulburd	$300-$750	unknown	19 rubies	19 jewels, thinnest Elgin, no two cases alike
1924	Gruen — Anniversary	$500-$2500	650	23 jewels being 2 diamonds, 21 rubies	12K gold bridges, Gruen pentagon cases
1925	Hamilton — Masterpiece	$300-$875	2600	23 rubies	

Table IX
Retail Prices ca. 1920

Type of Watch	Price	Examples
Premium grade, 18K case	$450	Premier Maximus, Edward Howard, Gruen Anniversary, C. H. Hulburd
Model T Ford, black	$300	
Extra quality grade, 18K case	$180	Riverside Maximus, Keystone Howard 23-jewel, Elgin 162
Top railroad grade, 14K case	$90	Elgin Veritas, Waltham Vanguard, Hamilton 946, Illinois Sangamo
Common rail grade, gold filled	$40	Crescent St., B. W. Raymond, A. Lincoln, Hamilton 992
Common 17-jewel, gold filled	$18	
Common 7-jewel, silver case	$7	
Inexpensive 7-jewel, nickel case	$5	N.Y. Standard, Ingersoll Trenton
Ingersoll dollar watch	$1	Ansonia, E. Ingraham, New Haven

less elegant appearing, movement to the Riverside Maximus. With its up/down indicator, the Premier Maximus was a uniquely ostentacious timepiece. After competition from the Edward Howard model, Waltham sought to maintain its position by raising the Premier Maximus to $400, adding a sterling silver box, and offering a Class A Kew Observatory rating (at a fee). By the time Gruen Anniversary watches were available at $500, the Premier Maximus was $750.

The Edward Howard was a natural reaction from Keystone Howard Company, whose pricing structure and reputation were based on uncompromised quality. Dealing solely in premium watches at premium prices, Howard needed a fleet-leading model comparable to the Premier Maximus, and the Edward Howard was a distinctive candidate. It was unique among presentation watches for having a frosted gilt finish, sapphire jeweling, and a free-sprung balance. Howard and Waltham could dispute which had the finest timepiece, but Howard had protected its integrity, plus that of its customers, who paid a notable price to become Howard watch owners.

By the 1920's, Elgin also decided to introduce a new premium timekeeper to support its reputation. Their high-grade convertible model was long discontinued and their 16-size grades 156 and 162 were outmoded by the trend for small watches. So in 1922 they unveiled the C. H. Hulburd model, named after their company president. The Hulburd sported the daring singularity of being a 19-jewel movement, compared to 23 jewels in all other prestige watches. Wadsworth cases on Hulburds were truly unique, no two being alike. Finally, this was probably the thinnest pocket watch made by Elgin, in keeping with 1920's fashion.

In 1924, 50th anniversary of the company, Gruen presented their Anniversary model. These normally had Gruen's pentagon style case, housing a spectacular appearing mechanism. Plates were 12K solid gold with a floral engraved pattern and bright ruby jewels. Typical of Gruen, movements were manufactured in their Swiss shops then adjusted and cased at Cincinnati. Starting at $500, they were among the most expensive of presentation timepieces and, whether considered American or Swiss, were intriguing watches.

Hamilton's Masterpiece model was the remaining watch of this group. In keeping with their reputation, Hamilton felt obliged to offer a premium class movement, choosing a special version of their 922 model. These were unusually accurate timekeepers, but were unique among presentation watches for having no unique characteristics. As with Hamilton's later marine chronometer, more emphasis was placed on functional quality than on eye-catching features. Combining the Hamilton name with a relatively low base price, the Masterpiece also had the largest sales volume of premium grade watches.

Because of their high prices, premium watches had limited over-the-counter sales potential. Frequently, they were purchased from group funds for honorary presentations. Their creators never expected mass market appeal, and netted little profit from them compared to receipts from commercial grade watches. Through superior timekeeping, 18K gold cases, and astounding prices, premium timepieces were intended to rebuild America's watchmaking image and fend off the Swiss wristwatch invasion. It was an unsuccessful effort, for wristwatches ruled the market by the late 1920's, dollar watches dominated sales, and American jeweled watch production had receded to 1890 levels of one million units per year.

REFERENCES

1. Edwin A. Battison, *Muskets to Mass Production* (The American Precision Museum, 1976), p. 31.
2. Thomas De Fazio, *A Note Concerning the Establishment of Railroad Watch Inspection in America* (BULLETIN, National Association of Watch and Clock Collectors, Inc., August 1975), p. 262.
3. M. R. McKinnie, *Webb C. Ball Watches* (BULLETIN, National Association of Watch and Clock Collectors, Inc., February 1970), p. 180.
4. Henry G. Abbott, *Watch Factories of America* (reprinted Adams Brown Co., Exeter, NH), p. 67.
5. George E. Townsend, *American Railroad Watches* (Alma, Michigan, George E. Townsend, 1977), p. 4.
6. James B. Morrow, *The Man Who Holds A Watch on 125,000 Miles of Railroad* (Alhambra, California, Meredith Publications), p. 5.
7. Ibid., p. 7.
8. Larry Treiman, *Railroad Watches and Time Service* (National Railway Bulletin, National Railway Historical Society, No. 1, 1976), p. 7.
9. McKinnie, p. 182.
10. Treiman, p. 8.
11. Morrow, p. 12.
12. Personal correspondence, J. E. Coleman to E. T. Fuller.
13. Treiman, p. 11.
14. Larry Treiman, *Webb C. Ball vs. Henry S. Montgomery . . . "a species of delirium?"* (BULLETIN, National Association of Watch and Clock Collectors, Inc., February 1976), p. 49.
15. Ibid., p. 50.
16. Generally, information on presentation watches has been extracted from: Eugene T. Fuller, *The Priceless Possession of a Few* (BULLETIN Supplement, National Association of Watch and Clock Collectors, Inc., Winter 1974).

Table X

Accuracy of Watches

Type of Watch	Average Error			
	sec/month	sec/week	sec/day	min/day
English pocket chronometer	30			
Premium grade lever	45			
Railroad grade 15J ca. 1865	—	60		
Railroad grade 21J ca. 1900	—	30		
Common jeweled lever	—	—	30	
Inexpensive jeweled lever	—	—	—	1
Dollar watch	—	—	—	3
Verge watch ca. 1800	—	—	—	15

VII
WATCHMAKING MACHINERY

Note: Illustrations in this chapter depict semi-automatic machinery that pre-dates introduction of fully automatic tools in the mid 1880's. Consequently, machines illustrated each require an operator.

There had been a number of watchmaking innovations before the American industry began. Most notable were invention of wheel cutting engines in the late 17th Century, eliminating hand filed geartrains, and late 18th Century introduction of specialized tools to manufacture ebauches in large quantities. The latter brought many new machines and standardization guidelines to watchmaking while posing little threat to handcraftsmen, since no finished parts were made. As horology approached the middle 19th Century several individuals in Europe and the United States saw that machinery could also make finished watches, with interchangeable parts. But machines for performing complicated and delicate finish work promised to be costly and were sternly resisted by European craftsmen who feared for their jobs. So during the 1840's watches were still finished by customary hand techniques. As the European industry grew it was not mechanical innovation that increased production but careful management of labor. Watchmaking tasks were subdivided among specialists to increase efficiency and gain higher quantity production from traditional hand methods.

Under traditional watchmaking, wheels and pinions were carefully, or not so carefully, cut on engines little changed

142. Punching wheel blanks from strip brass.

141. Swiss watchmaking in the traditional cottage craft manner.

in 150 years, then matched to their particular set of watchplates. Balances and escapements were laboriously constructed by hand using simple fixtures, then mated to the watch. Hand files were among the most important tools and the watchmaker's lathe was still a simple set of turns. Turns were merely a frame in which the workpiece was turned between dead centers by an outside power source, usually a bow. The watchmaker had to bow the workpiece with one hand while holding the graving tool in the other, and thus were made many of the finest watches and chronometers ever produced. The idea was just gaining popularity that a foot treadle could power the lathe, freeing one hand and maintaining constant rotation in one direction, but this sort of gimmickry was resisted by traditional watchmakers. Another new idea at the time was the mandrel lathe, which held watchplates in a rotating headstock powered by foot treadle or hand crank. It also controlled the graving tool with a compound slide rest, and had additional accessories that made it a sort of universal watchmaker's tool. Unfortunately, mandrel lathes often sat as idle window displays since few jewelers understood their

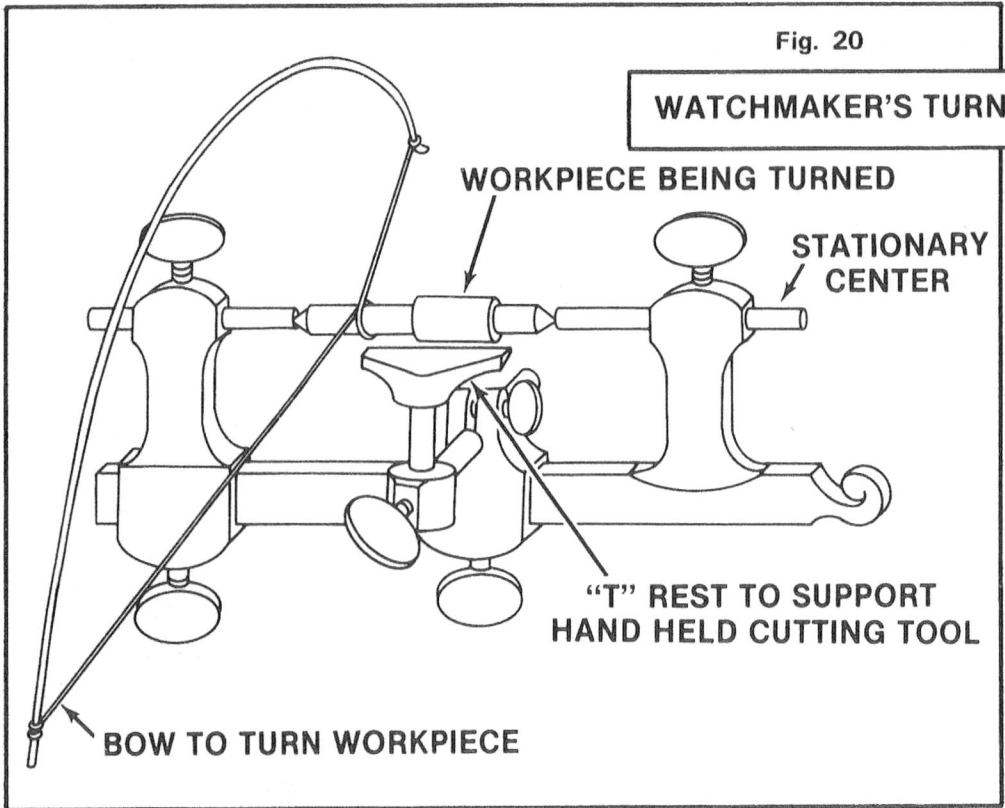

Fig. 20

WATCHMAKER'S TURNS

WORKPIECE BEING TURNED

STATIONARY CENTER

"T" REST TO SUPPORT HAND HELD CUTTING TOOL

BOW TO TURN WORKPIECE

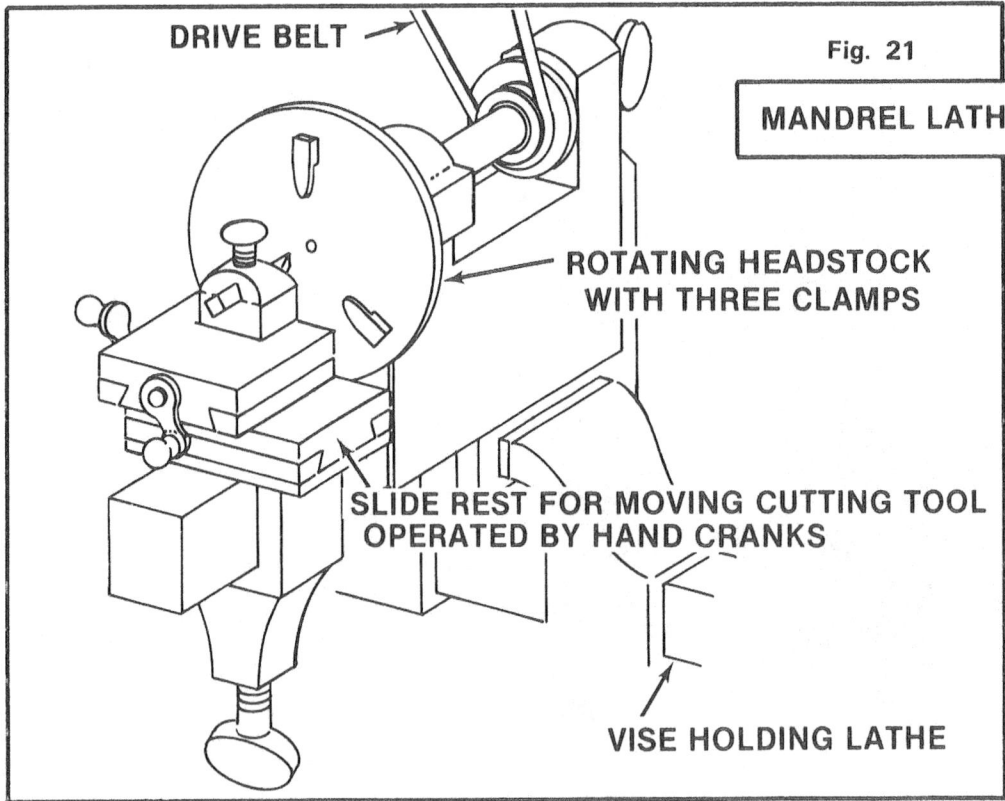

DRIVE BELT

Fig. 21

MANDREL LATHE

ROTATING HEADSTOCK WITH THREE CLAMPS

SLIDE REST FOR MOVING CUTTING TOOL OPERATED BY HAND CRANKS

VISE HOLDING LATHE

143. The Elgin train department.

use. Watchmaking in 1840 was still the watchcraft of old, with watch parts custom-made by individuals hunched over simple hand tools.

In contrast, the United States had several examples of mass production with interchangeable finished parts by 1840, the Connecticut clock industry making wooden works shelf clocks, and the Springfield Armory making rifles. Similar methods were used for American watchmaking and, not surprisingly, some watchmaking pioneers spent part of their early careers at the Springfield Armory. Even Aaron Dennison visited there to observe their operations. Machines were the heart of such mass production efforts and mechanics who designed them were the key people. In its infancy the industry did not know how to make, nor could it afford, sophisticated machinery. Only limited capital was available and the pioneers had to demonstrate their potential in a somewhat primitive manner. Early machines were consequently crude by later standards and each one required constant individual attention from an operator. They still performed an adequate job for a modest company and even as the industry advanced, early tools were rarely discarded. They were sold to another company or updated with new inventions in lieu of building costly new machines from scratch. A long-standing factory like Waltham could thus be found with tools built on durable lathe beds that were nearly a century old. The difficulty and expense of developing machinery prevented the early industry from making certain watch parts at all. Dials, hands, springs, jewels, pinion wire, and balance wheels presented problems for years and were imported by many companies into the 1880's. Even as companies began making these parts it was common to use hand methods and employ English or Swiss workmen, especially for dial painting, gilding, and jewel work. The cost of such manual labor prompted the Tremont Watch Co. to build their trains, escapements and balances in Switzerland.

It can be seen how slowly these problems were solved by reviewing the development of hairspring making. Imported balance springs were one of the main drawbacks in making precision watches since they were nearly all soft springs. They were hardened simply by cold rolling and as a result changed their properties over the long run. Heat treat hardening was required and about 1860 James Bottom of New York developed a furnace process for hardening and tempering springs, which he kept so secret even his apprentices knew nothing of how it was done. (In 1850 he had patented "Bottom's lathe," likely the first foot-treadle lathe in America, and did a fair business with them in the jewelry trade.)

His first springmaking efforts involved winding the soft spring wire into a heat treating box along with a soft piece of thin mainspring to achieve coil spacing, thus tempering only one hairspring at a time. Later he developed a method of winding three hairsprings together, each one spacing the coils of the others. For a number of years Bottom was the only source of heat treated springs, providing E. Howard and the United States Watch Co. of Marion all their hairsprings, as well as selling to the general trade. During this period John Logan was making hairsprings for the Boston Watch Company, where Charles Van der Woerd had also made some improvements in the production of hardened springs. In 1863 Logan joined E. Howard & Co. and, along with William Todd, perfected more efficient techniques for tempering springs, from whence he produced all Howard's springs till leaving in 1870. After several ventures plus another term of employment at Waltham, Logan opened his own spring company in 1877 where he invented a method for hardening and tempering Breguet overcoil hairsprings. In addition he made mainsprings and springs for other industries. For many years the Waltham Co. bought his entire output of Breguet springs, although he sold flat hairsprings to other companies since the Waltham plant made their own. Many companies were too small to have their own springmaking department and until the success of Logan's process there were few domestic sources of balance springs.

Workhorses of the watch companies were arbor and staff lathes, wheel and pinion cutters, screw-making machines, jewel lathes, punch presses, plate lathes, drilling and tapping machines, and a multitude of small tools and gauges. Incorporation of a watch company was often followed by several years of machinery making before any watches were produced, and up to 1/10 the employees of a well developed company would be in the tool department. The expense of accumulating machinery was one of the major stumbling blocks in starting a watch factory and new companies often started by purchasing the machinery of some defunct firm. Ultimately, their watches could only be as good as their machinery, and going into business on bad machines was inevitably followed by going out of business on bad machines. Many entrepreneurs underestimated the investment required for watchmaking and sowed the seeds

144. Wheel cutting machine — a stack of wheel blanks is being milled by a fly-cutter. The indexing head is under the large circular base.

of their own demise when they purchased second-hand tools. These had often been specifically designed to produce an obsolete watch, required highly trained operators to run smoothly, and even at their best could not compete with efficient new machinery being used by large successful companies.

Nearly all early watch companies acquired machinery from the remains of previous firms, with results that varied from excellent to dismal. Edward Howard successfully started a watchmaking business with material and tools from the old Boston Watch Company. The American Watch Co. established its ¾-plate department by purchasing the Nashua Watch Co. machinery. All the companies in Lancaster, PA, built upon physical remains of their predecessors, from Adams & Perry to the Lancaster Watch Company to Keystone Standard to Hamilton, which also bought old Aurora Watch Co. machines. The South Bend Watch Co. was formed by shipping in the Columbus Watch Co. complete. In 1930 Russia bought Hampden Watch Co. and the Ansonia Clock Company, moving them in their entirety to establish an industry in Russia. On the other hand the New Jersey twins, Newark and U.S. Marion, created a wave of itinerant machinery that seldom saw a sunny day. Remains of the Newark tools went to Chicago as the Cornell Watch Company, then on to San Francisco as the west coast edition of Cornell, back to Chicago as the Western Watch Company, then off to the Independent Watch Co. of Fredonia, New York. Meanwhile U.S. Marion machinery was being disseminated to Ezra Bowman in Lancaster, PA (who later sold it to J. P. Stevens of Atlanta) and others. The Union Watch Co. of Fitchburg, MA, got some of it, as did the Auburndale Watch Company. Other Marion machines went to the Independent Watch Co. to join old Newark tools. Independent's collection of ancient New Jersey machinery finally served to form the Peoria Watch Co. in Illinois, with the result that a Peoria watch of 1890 looked like a Newark watch of 1870. In its lifetime, this mass of Marion and Newark machinery produced fewer watches than Waltham made in the year 1890 alone.

In order to pioneer mass production of interchangeable watch parts, Dennison and Howard had to pioneer watch machinery suitable for factory use. It was no surprise that the first and most basic improvements were on the lathe, one of the most fundamental tools in any industry. The most important features were developed by Charles S. Moseley, who joined Dennison and Howard at the Roxbury factory in 1852, before watches were put on the market. It was apparent to him that the mandrel lathe offered the most promise with its live headstock and compound slide rest. But it came with only a face plate having three clamps, or a wax chuck in which the workpiece was secured by melting hard wax around it, neither of which looked useful for factory production. Moseley noticed that there existed, separately from watchmaking, a small hand tool with a spring chuck mounted on a hollow wood handle, in which various tool bits were stored. The tool bits could be inserted into the chuck, which was then screwed tight to hold them. He applied this approach to the mandrel lathe by placing a similar chuck into a hollow headstock with a hand wheel at the back for tightening the chuck. Thus he constructed the first collet lathes, designed for working on both watchplates and wire rods. Next he put adjustable stops on the slide rest so that it could be repeatedly cranked to the same setting with precision. Lathes of this sort were used by the Boston Watch Co. and all other companies subsequently. The hand wheel for closing collets was replaced by a foot pedal, cranks on the slide rest were replaced by hand levers for quicker operation and additional pedals were introduced

for starting and stopping the lathe, so that there was soon a tool for mass production.[1]

Such lathes were suitable for turning plates and roughing arbors, but it was found that for finishing arbors and staffs, supported at only one end by the collet, the cutting forces deflected the arbor being cut and resulted in a defective part. Consequently, through the 1880's, arbors had to be turned between dead centers with a lathe dog clipped to the arbor. The lathe dog was then powered by a belt driven whirl around one of the stationary centers to rotate the arbor in the lathe. With the lathe dog obstructing one end of the arbor, work could only be performed on the other end. Moreover the other end could not be machined down too far or it would not be strong enough to support the cutting forces when the unfinished portion was machined. Therefore the arbor had to be repeatedly redogged and put into another lathe so that the two ends were machined down evenly, resulting in more than ten lathe operations on a complex part like a balance staff. To work such factory lathes an operator had to mount the workpiece in the lathe, putting lathe dogs on arbors, then close the chuck or bring in the centers, turn on the lathe, operate handles on the slide rest to perform the cutting, stop the lathe and finally remove the piece and take off the lathe dog. Both arbors and plates could be machined this way but much effort was required. Plates were also complex, having numerous recesses and countersinks, plus holes both plain and threaded. Many lathes were required and each one needed an operator constantly working over it, representing a large investment in machinery and wages. In addition, plate lathes had a problem since collets were closed down onto the part by pulling back into a tapered recess in the headstock. If plates varied in diameter, collets would pull in correspondingly different amounts and a given setting on the slide rest would machine the plates to varying thicknesses. To overcome this, Moseley arranged for collets on plate lathes to remain stationary while the headstocks moved forward to close them, always maintaining the same distance between the plate and slide rest. Over the years many such difficulties were ironed out but the main features remained, easy foot pedal operation and hand lever tool movement. Precision stops on the slide rest resulted in all parts being machined identically, and trained operators could produce many parts each day.

Though arbors were still turned on dead center lathes, the 1860's and 70's saw increasingly automatic functions. The slide rest, which held the cutting tool, was operated by cams, while the lathe automatically started, stopped and opened the chuck or drew back the centers to release the workpiece. Operators had only to insert new blanks and put lathe dogs on arbors. In early years arbor blanks were cut from wire, roughed out in a collet lathe and pointed on one end, then reversed and pointed on the other end so they could be turned between dead centers. About 1880 Charles Vander Woerd invented a machine that was really two collet lathes facing each other. Wire was roughed and pointed on one end in the first lathe, then automatically cut off and transferred across to the opposite facing lathe with the unfinished end facing outward. That end was pointed and the blank was deposited into a bin separate from machining chips, since separating small parts from chips had been a problem for years. One operator could oversee 6 or 8 such machines, each digesting lengths of wire several feet long and converting them to blanks ready for dead center arbor lathes. Arbor lathes were next improved so that a driver would automatically grasp the arbor as the centers closed onto it, eliminating the lathe dog and leaving operators only the job of loading and unloading machines. Automatic drivers also supported arbors more firmly, al-

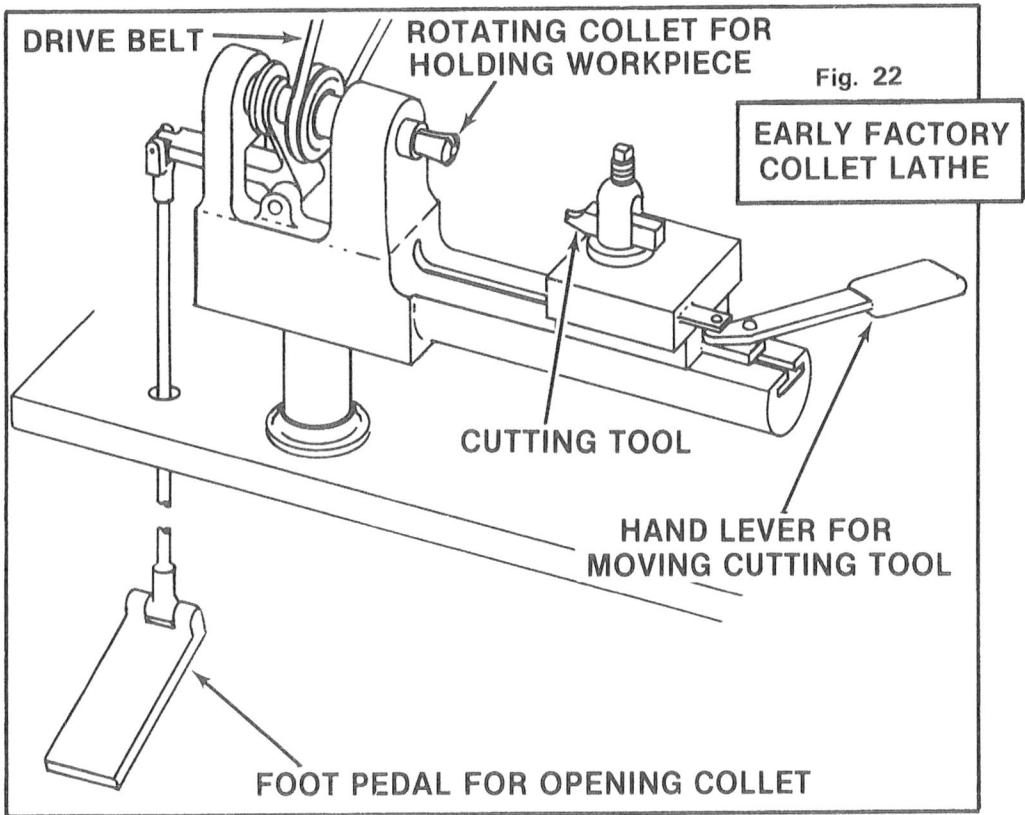

DRIVE BELT — ROTATING COLLET FOR HOLDING WORKPIECE

Fig. 22

EARLY FACTORY COLLET LATHE

CUTTING TOOL

HAND LEVER FOR MOVING CUTTING TOOL

FOOT PEDAL FOR OPENING COLLET

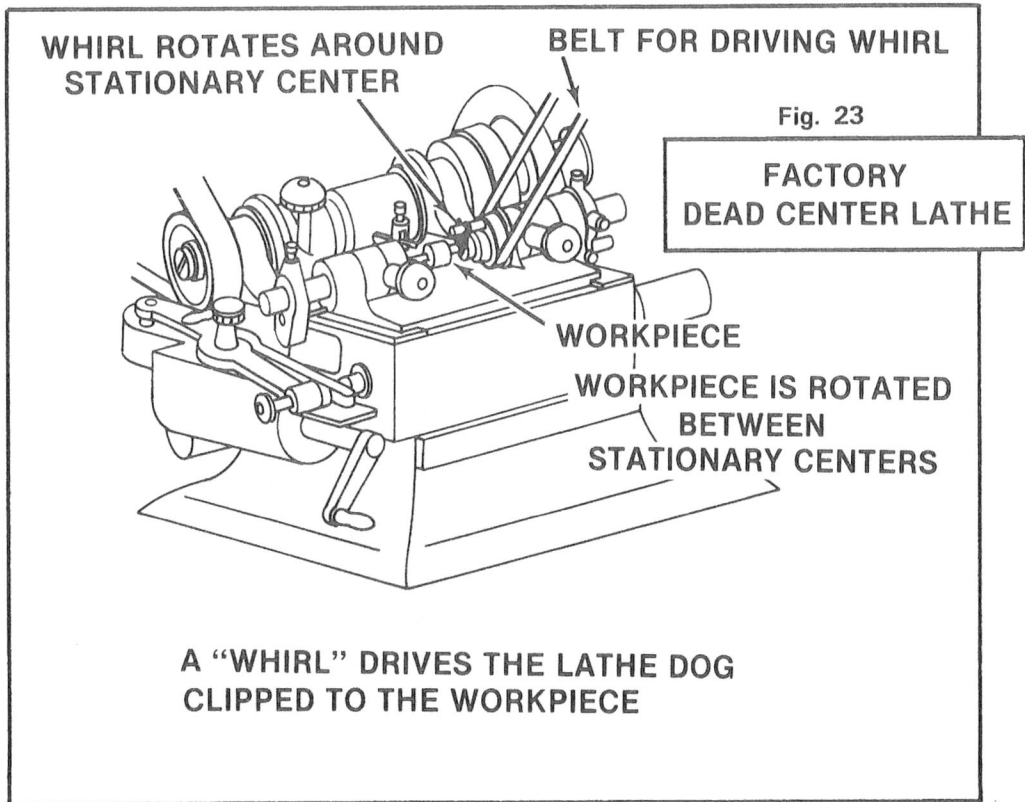

WHIRL ROTATES AROUND STATIONARY CENTER

BELT FOR DRIVING WHIRL

Fig. 23

FACTORY DEAD CENTER LATHE

WORKPIECE

WORKPIECE IS ROTATED BETWEEN STATIONARY CENTERS

A "WHIRL" DRIVES THE LATHE DOG CLIPPED TO THE WORKPIECE

lowing heavier cuts and reducing the number of lathe operations required to make finished pieces. Manufacturing steps and time consuming hand manipulations were thus eliminated, progressively leaving operators only the job of feeding machines. In the next logical move Duane Church invented a series of transfer arms which removed workpieces from one lathe and placed them into the next, ready for work to commence. With this development, a row of finishing lathes could be placed behind a roughing lathe and they could all perform their operations in unison, stopping while transfer arms passed unfinished parts along to the next lathe. A bank of five machines could complete finished parts and one operator could care for several banks of lathes, feeding lengths of wire into one end and collecting finished arbors at the other.

These factory lathes were heavy steel machines about the size of typewriters. To maintain fine tolerances required for watch parts, they were solidly built so that they could not warp, deflect, or otherwise come out of adjustment. They had a complex series of cams, levers and shafts to perform all the automatic functions such as turning the part, operating the cutting tool, opening and closing the chuck, starting and stopping the lathe, etc., and separate drive belts to power these. While they produced thousands of parts daily to earn their keep, they were highly complicated and a bank of five such machines was a considerable expense to build and maintain. As a result, Mr. Church made the final improvement around 1895, eliminating the need for five lathes to make one part. He overcame the problem of turning fine dimensions on a collet lathe by introducing numerous automatic cutters, so that enough light cuts could be made to complete a finished part. The lathe accepted four-foot lengths of raw wire stock and turned out finished balance staffs with greater accuracy than before since the staffs did not have to be transferred from one machine to the next. In addition, if one such lathe broke down it did not interrupt the work of four other machines, as would happen in a bank of lathes.

Duane Church incorporated many such improvements in the Waltham plant. He designed banks of plate turning lathes with transfer arms so that individual plates would progress down the row and be finished on both sides. In his machines the transfer arms and slide rests were moved up against hard steel stops by compressed air, providing rapid motion with high accuracy. Stops could be reset and control cams changed so that machines were used to make a number of different parts. He also improved pinion and wheel cutters, screw making machines and others, generally preferring the approach that one machine should make finished parts from rough stock without transferring the workpiece to another machine.

Wheel and pinion cutters developed gradually over the

years, much like lathes. The original wheel cutting engine, invented in England around 1675, had seen no drastic improvement when Dennison used an imported engine at the Roxbury factory in 1850. Machines operating on similar principles were used in the factory into the 1860's, though improvements had made them almost automatic. Pinions were made from one-foot lengths of pinion wire, imported from England, that had been drawn through a progressive set of dies to form rough pinion leaves, or teeth. Pieces were chopped off to be roughed and pointed in roughing lathes then finished turned in arbor lathes. At that point the pinion leaves were still in the rough and the arbor had to be completed on a cutting engine. Eventually Charles Vander Woerd, in the Nashua department, produced an automatic pinion cutter. It had been found poor practice to have one cutter form the entire tooth from scratch, so Woerd mounted the pinion to be cut under a turret containing three separate cutters. The pinion traversed horizontally back and forth under one cutter, indexing between each pass, and when all the teeth had been machined by that cutter the turret rotated to bring the next cutter into operation. With such a machine pinion leaves could be cut into arbors made from plain round stock and the company was no longer dependent upon imported pinion wire. Not long afterward Ambrose Webster, of Waltham's fullplate department, made a pinion cutter on a slightly different plan. Instead of three cutters on separate spindles, he mounted the cutters on one shaft, which moved in its endwise direc-

145. Waltham pinion room.

146. and 147. Two versions of Waltham turret-type cutters for machining escape wheel teeth.

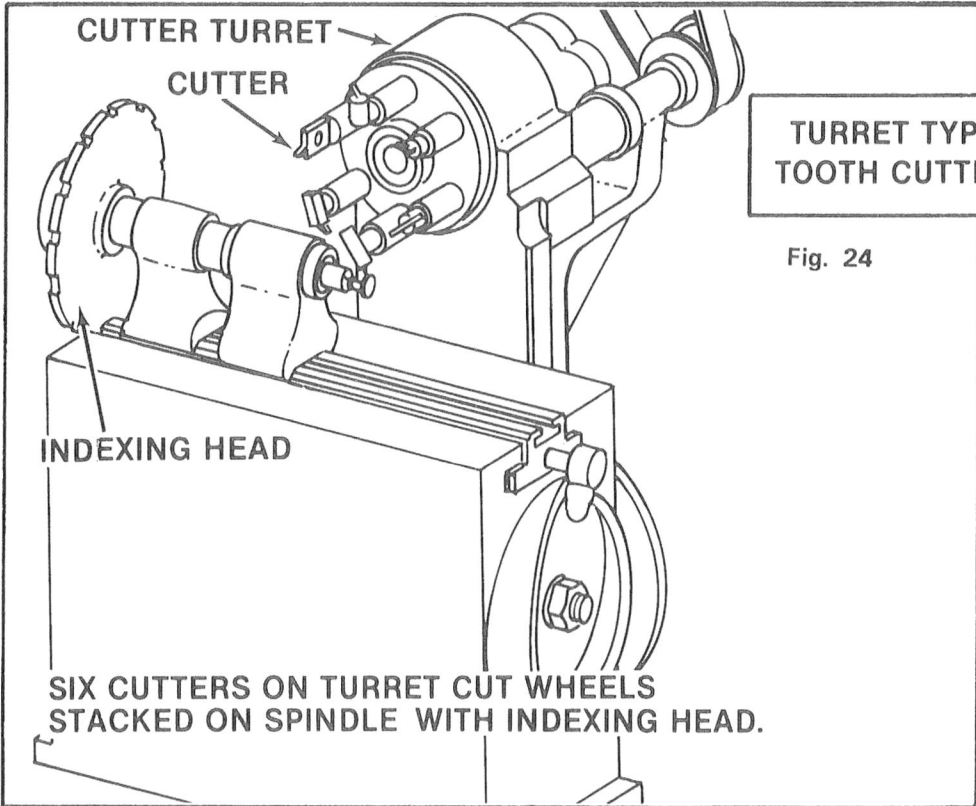

CUTTER TURRET

CUTTER

INDEXING HEAD

TURRET TYPE
TOOTH CUTTER

Fig. 24

SIX CUTTERS ON TURRET CUT WHEELS
STACKED ON SPINDLE WITH INDEXING HEAD.

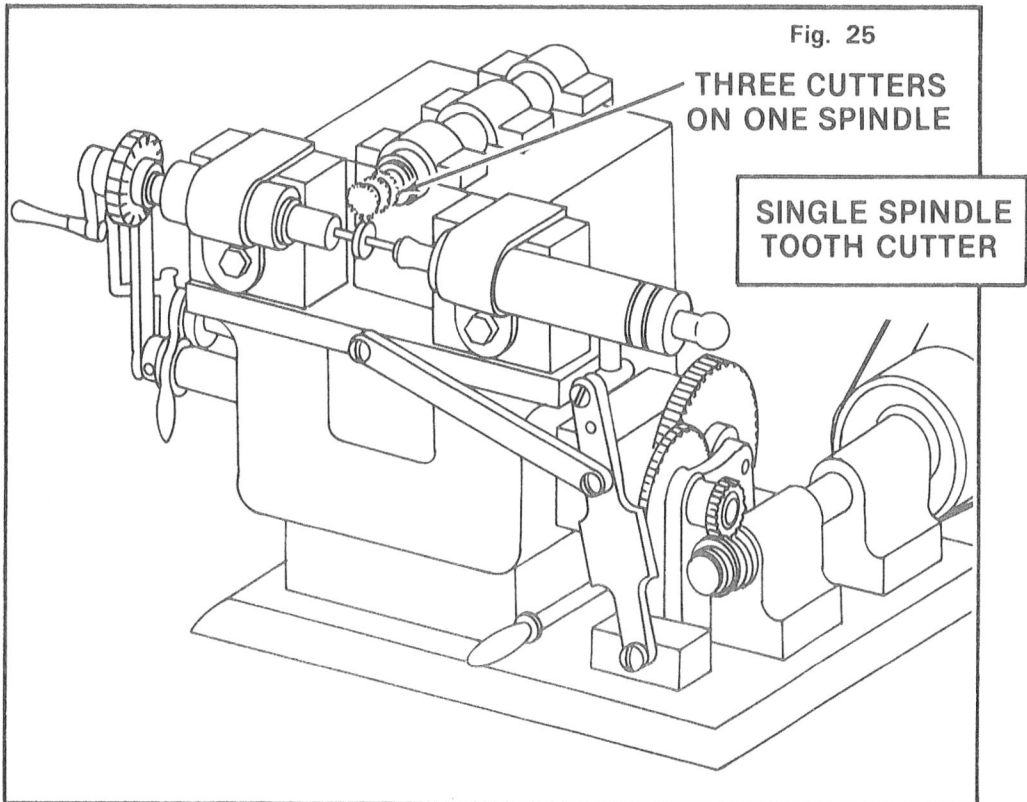

Fig. 25

THREE CUTTERS
ON ONE SPINDLE

SINGLE SPINDLE
TOOTH CUTTER

tion to bring the separate cutters into operation. This machine was easier to adjust than Woerd's turret-type cutter, so became standard throughout the plant thereafter. Both types left their operator only the job of loading and unloading blanks, but still required an operator for every machine. Eventually a group of 8 cutters was set on a rotating table, with magazines loading blanks automatically while finished parts dropped into a receptacle.

Wheel cutting, like pinion cutting, was gradually automated. Operators originally performed all indexing and changing of cutters by hand, but Mr. Webster invented semi-automatic machines as early as 1865, which became nearly automatic in the early 1870's. Machines were designed to cut wheels in stacks, producing large quantities with ease. Escape wheels were more complex, requiring up to 8 cutters, and followed the idea of having a turret with

148. Waltham screw making department.

cutters on separate spindles. Webster and Woerd made machines with 6 cutters on the turret and by 1883 these had become completely automatic. Steel escape wheels became a factor in the 1890's since they were required in railroad watches and were harder to cut than brass wheels. Consequently a saw cutter was used for gross metal removal in making steel wheels and milling cutters finished the final contours.

Screwmaking was also important since there were so many used in watches. Early screw lathes were simple hand lathes with a double slide rest and double tailstock. Wire rod was fed through the hollow headstock and held in a collet. The first tailstock gauged how much wire was fed through and the first slide rest machined the shank of the screw. The second tailstock held a thread die and the lathe was turned by hand while the die was run onto the screw shank. Finally the second cutter almost parted the screw off but left a tiny bit of metal holding it. The lathe was again turned by hand to feed the screw into a plate which held two rows of screws and the screw was then twisted off. When the plate was filled a saw was run down the two rows to form slots in the screw heads. Automatic screw machines were first designed by Charles Vander Woerd in 1871 to make jewel screws, since these were quite small, and a machine for larger screws was made about four years later. Woerd's machine, which worked on one screw at a time performing operations separately and in succession, was the tool which brought Waltham so much attention at the Centennial Exposition in 1876. It was also incredibly complicated, and a simpler version was invented which held several screws in separate spindles on one turret. Each spindle was worked on by its own individual tool, and when all operations were complete the turret was indexed a par-

149. Ladies at jewel making lathes.

tial turn to bring each spindle around to the next tool. Such a machine was capable of making 50,000 screws per day and one operator could attend six machines. Because of its efficiency this "screw machine" concept is still used to make small round parts in large quantities. The other machine common today for making such parts is the automatic collet lathe with multiple cutters, a derivative of Duane Church's arbor lathe.

Damaskeening was first practiced in America by the U.S. Watch Company of Marion, which secured a one-year contract with a Mr. Wilmont from St. Imier, Switzerland, to get started. After that year, Mr. Wilmont worked for Waltham and then Elgin before returning to Switzerland. The process involved charging a rotating wood dowel, or disk, with polishing compound, then moving it over the surface of the watchplate (or moving the plate under the ro-

150. Machine for gauging strength of hairsprings.

tating dowel, or a combination of both). Wood disks gave way to ivory and other materials, and machinery was programmed to move watchplates while the rotating tool remained in a single location or moved to and fro in one direction. The small tip of the tool buffed a textured pattern into the polished plate, plates being either gilded or nickel. Combining patterns of spots and lines yielded a virtually infinite number of designs, using buffing tools of various diameters and complicated programs for rotating and translating watchplates. Some damaskeening machines, at least through the 1870's, allowed the operator to vary designs at will, so that on some watches, such as Waltham's 21-jewel model 72, there may be no two movements with identical patterns.

Dial making presented technical difficulties, so that through the 1880's many companies imported dials, as they did balance wheels and balance springs. Even as factories began making their own, Swiss and English dial makers were often employed. To manufacture dials, a round copper disk, having the feet brazed on, was covered on both sides with a paste of fine white enamel powder. This was placed on a thin steel pallet at the mouth of a furnace to fuse the enamel, after which the rough surface had to be ground smooth and polished. It was returned to the furnace for a second firing then sent to the dial painters. Painters laid out the chapter rings, numerals, minute divisions, and company name, using black enamel paint, hand applied with camel hair brushes. Eventually, painted features were transferred onto dials by means of rubber printing blocks. In either event, the dial was then fired a third time to fuse the enamel lettering. Because white enamel base was baked onto both sides of the dial, it could tolerate these refirings without warpage or cracking. Where the seconds bit was to appear, the enamel was ground away from both sides and the underlying copper etched out with acid. After the edge of the hole was polished, a seconds dial, prepared by the above process, was soldered in place.

Many years passed before efficient methods were developed for all the varied aspects of watchmaking. Balance wheel assemblies were complex to make, consisting of a temperature compensated balance wheel, hairspring with collet and stud, balance screws, balance staff, roller table and roller jewel. Numerous machines and hundreds of operations were required, plus assembly and adjustment of the final product. The problem of matching balance wheels with balance springs was always a bit complicated, especially by the time adjusting and temperature compensation were looked after. When companies gathered enough experience they could weigh each balance and gauge each spring against a master, noting its strength. These could then be matched in whatever manner had been found to give best

152. Firing dials.

results, and regulated to vibrate with the correct frequency. Other processes presented problems such as jewel making, grinding, lapping, and polishing steel parts, especially ones that were not round. Each problem was eventually solved and the industry went on to make all watch parts in quantity and at low cost.

Watch company buildings shared a number of common characteristics. As substantial manufactories, they were usually brick buildings several stories high, with narrow wings having the maximum possible number of windows. The tool department and heavy plate punching presses were in the bottom floor, where their weight could easily be supported. Size and number of machines then decreased as one ascended to the top floor. Boilers and steam engines were often located in a back wing. Boilers could supply both building heat and steam for an engine. In early years of the industry, engines under 50 horsepower provided energy to an entire watchmaking operation. Furnaces for baking dials and heat-treating steel parts could be located near the boilers, to share a common chimney, or might have their own area at a large factory.

Upstairs were machine rooms, known as automatic rooms in later years. Low-speed shafts were supported from the rafters, driving through leather belts to countershafts on the workbenches. Countershafts had pulleys of appropriate sizes to drive machines located along the benches. Narrow wings of the building usually had benches all along their outer walls, adjacent to windows, plus a row of benches down the center. This layout provided ample light for small-part work, important for both machine rooms and assembly or adjusting areas. Since human eyesight is keenest in the wavebands of natural sunlight, windows were essential even after gas and electric lighting were common. Through the 1890's, pipes and rubber hoses for compressed air were added, since some machines had certain of their

151. Balance making room.

functions powered by air. Machine rooms thus played an automated symphony, as shafting rumbled overhead, countershafts whirred, machines clicked and clattered, and air hoses hissed. A few caretakers attended machines, seated on stools rolling on tracks along the benches, feeding raw stock into machines and collecting completed parts. A maintenance crew set up and adjusted machines, replaced worn cutters, unsnagged hung machines, and made repairs.

On the floor above, assemblers staked wheels onto pinions, balances onto staffs, set jewels into plates and pallets, ran trains, and assembled movements. Movements would be run prior to final finishing, held together by rough screws so that finished screws would not be marred or tarnished. After gilding and finishing, movements were then timed and adjusted. Departments were organized by the parts they made or service they provided: eg. plate making, wheel making, wheel cutting, pinion roughing, pinion finishing, flat steel making, gilding, adjusting, etc. In a large factory, finishing a thousand watches per day, a movement could be one year creeping from department to department as its parts were roughed, assembled, and finished. When parts were sufficiently complete to be serial numbered with their mating pieces, the numbered set moved through the building in a compartmented traveling tray. As it neared a finished state, the movement could then be assembled and housed in a serial numbered box, in which it was ultimately shipped from the factory.

It can be seen that machine designers were masters of

the watch factories, but this was not recognized as early as one might expect. After working as a machinist in several other businesses including the Springfield Armory, Ambrose Webster joined the Waltham factory in 1857 as their first general machinist with no watchmaking responsibility. At the time, management saw watchmakers as the chief people in the plant and resisted efforts on the part of Webster to promote machine methods. Stubbornly, Webster expanded the tool department and built semi-automatic machinery, even against the will of the company. He also organized the factory into one measuring system, where there previously had been nine or more, and did a great deal to standardize lathe designs so that lathe tools and accessories would be interchangeable. As chief mechanic of the fullplate department he made tremendous contributions to the automation of watch manufacture, which helped build Waltham's strong position. As the industry spread it was not watchmakers who went forth as apostles, but watch machine makers. Pioneer mechanics such as Moseley, Hunter, Bartlett, Stratton and Woerd plus later men like Duane Church and A. E. Marsh developed the American system of watchmaking.

From a modest beginning, factory watchmaking techniques grew steadily into a full scale industry. Dennison and Howard began the system with such helpers as Nelson Stratton and C. S. Moseley. When their Boston Watch Co. failed in 1857 Dennison stayed briefly with the new owners then started the Tremont Watch Company while Howard

153. Grand view down a wing of the Waltham factory.

154. Final movement assembly at Elgin.

155. Engine room, with sun roof, at Waltham.

went back to his original Roxbury factory. Stratton, Moseley, Woerd and J. H. Gerry went to Nashua and in their abortive attempt to start a watch company, built the finest machinery to date. Moseley, George Hunter and P. S. Bartlett later went to Elgin, Illinois along with several other Waltham mechanics, and a number of these men went on to the Illinois Watch Co. in Springfield. Hunter had built a number of Webster's machines at Waltham and was long one of the chief machine designers at Elgin. Bartlett had been in charge of the plate and screw department at Waltham and the company came out with an 18-size movement in 1859 which bore his name, as well as a 10-size ladies' movement in 1861. (He came from one of the oldest puritan families in New England and his great uncle was a signer of the Declaration.) James Gerry returned to Waltham for a time then went to U.S. Marion as a tool maker and also worked for the New York Watch Co. and E. Howard & Co. where he invented Howard's stem winding mechanism. The Newark Watch Co. started with N. B. Sherwood who had invented a number of self-measuring and automatic machines for Howard. Charles Vander Woerd was superintendent of the Waltham plant when he left in 1882 to establish the Waltham Watch Tool Company, and later started the United States Watch Co. of Waltham, designing their first watch. These men and their followers continually advanced watch manufacturing techniques. As industry competition increased through the 1880's, there was renewed emphasis on improving machinery in order to reduce watch prices. Designers created fully automatic machines where previous ones had been semi-automatic. By 1900 watchmaking was one of the most sophisticated branches of industry.

As the watch industry expanded, machinery making went well beyond the walls of watch factories. Watch machine designers began their own tool companies, spreading the conveniences of their machinery to other factories and individuals. While developing factory lathes, Charles Moseley also produced a small lathe which could serve repairmen. During his stay at Elgin, he joined his brother Horace in forming a company to produce jeweler's bench lathes, which Horace continued for years. Ambrose Webster established a tool factory in 1876, along with John Whitcomb, called the American Watch Tool Company. They marketed Webster's standard lathe bed, with interchangeable accessories, and the "W-W" lathe is still the standard design on watch and jewelery makers' benches. In 1878 that company equipped the new Waterbury Watch Co. to make 1000 watches per day. While working on the Waterbury project Webster met Mr. Woodbury of Seth Thomas, inducing him into the watch business and later selling Seth Thomas considerable amounts of machinery. The American Watch Tool Co. also equipped the New Haven Watch Company, helped plan the Trenton Watch Company, and over the years was instrumental in supplying planning and machinery for the Cheshire, Columbus, Aurora and Hampden companies. Likewise, after leaving Waltham in 1882, Charles V. Woerd began the Waltham Watch Tool Company, one of many such organizations to spring up in Waltham. Jonas Hall, who once worked at Waltham, began the J. G. Hall Mfg. Company in Roxbury, Vermont. He made numerous hand tools, but most importantly, innovated and spread use of the watchmaker's staking tool throughout the retail trade. Companies like these not only advanced watchmaking, they made their sophisticated tools available to other industries. Screw machines and automatic lathes began serving all branches of mechanical manufacture. Capability to produce interchangeable parts led to adoption of national and world standards of thread gauges, screw sizes and other uniform systems. The obvious convenience of such standards helped coordinate the industrial standard of living we all enjoy today.

REFERENCES

1. Information on watch machinery at Waltham has been summarized from: E. A. Marsh, *Watches by Automatic Machinery at Waltham* (reprinted Exeter, NH, Adams Brown Co., 1968).

All companies had similar machines, usually purchased from watch machine companies. Waltham and Elgin made much of their own machinery and small tools. Designers at Elgin apparently preferred a carrousel approach, with a rotating platform indexing around to present parts in process to separate tools stationed around it. On the other hand, Waltham machines often kept the part in one location while different tools were brought around to work on it. Variations on these concepts today form the basis for most high-speed production equipment.

SECTION 3
OUT OF THE ORDINARY

VIII
VARIATIONS AND ODDBALLS

The American watchmaking industry developed a fairly uniform product, almost standard in general description. Making the average American watch became the recipe to success, or so hoped numerous businessmen of the day. There were also those who were motivated to make something out of the ordinary. Once the credibility of American watches was established it became a gamble to manufacture a new or different product. Such a venture entailed not only the usual business risks but required the added complication of convincing people they needed a new kind of watch which they had previously managed to live without. There were a number of new ideas that made sense, such as inexpensive watches that brought timepieces to millions of people who had previously managed to live without any watch at all. In this instance the unique nature of the mechanism made it cheaper to produce, thereby providing a service to the public as well as profits to the inventor. Other unusual watches were likewise pure in their intentions, often representing honest searching for better timekeeping, lower cost, greater durability or an equally worthwhile cause. In other cases the unusual nature of the watch was merely to attract the attention of a buyer. Whatever the motives, such watches were made in sufficient quantity to deserve mention.

The first oddball watch in this country was the Howard Davis & Dennison. These were the first 17 watches produced by the Warren Mfg. Co. in 1852, with the peculiarity that they ran for eight days. This idea was tried occasionally throughout watchmaking but rarely in America. The only other significant attempts were large watches made by Waltham and Elgin to be used as table clocks or automobile clocks, but not as pocket timepieces. The Howard Davis & Dennison was more of a false start than a variation since the industry did not even exist then let alone have a standard product. It was Dennison's thought that an eight-day watch would be popular, but a survey of local jewelers indicated that he would have better luck selling them a one-day model. Dennison had first tried making the watch run for eight days on one mainspring but that was not successful and the Howard Davis & Dennison used two mainsprings in tandem. Material for a hundred was started but when market prospects appeared low, and technical difficulties with them were great, only 17 were completed and presented to investors rather than offered for sale. After that Dennison marketed what became the model 1857 Waltham and the American watch industry was born.

Of all variations from the average watch, the oldest and most common are probably the timer and chronograph. The two terms do not mean the same thing. A timer is a watch used solely for timing intervals. It therefore has a sweep hand which can be started, stopped and usually returned to zero, but there is no hour or minute hand, and when the timer is stopped the entire mechanism stops running. (As usual there is at least one exception to this rule.) A chronograph is a normal timekeeper, therefore with an hour and

156. Dial of Waltham Chronodrometer #13,744 ca. 1859 — a sweep hand rotates every four minutes and the seconds hand rotates once in four seconds. An hour and minute hand, with time dial, may be used for time of day.

157. Movement of Chronodrometer #13,744.

158. Dial of Auburndale Timer ca. 1880 — sweep hand rotates once each minute and the seconds hand once each second, jumping quarter seconds. The recorder hand records elapsed minutes, rotating counterclockwise.

minute hand, plus a sweep hand which can be started, stopped and returned to zero. Since a chronograph has to maintain the correct time of day the mechanism must continue running when the sweep hand is stopped. Some device is required that can drive the sweep hand but disconnect it when desired. It can be seen that a chronograph is a conventional timepiece which can also be used as a timer.

The first timer in United States watchmaking was the Chronodrometer produced by Appleton Tracy and Co. in 1859. This was based on Dennison's model 1857 movement, with a stop mechanism operated by a button on the side of the case. A sweep hand revolved once every four minutes and a small second hand at the usual six o'clock position revolved once every four seconds, neither of which could be returned to zero. At 12 o'clock was a small dial with an hour and minute hand; so if these were set to the correct time of day and the watch was left running, the Chronodrometer could be carried as a normal watch. Correct time of day was interrupted when the watch was used as a timer, requiring the hour and minute hands to be reset afterward. Since this amounted to an inconvenient situation, few Chronodrometers were sold, rapidly ending one of the earliest attempts by the industry to expand its product line.

A timer wasn't made again till 1876 when the Auburndale Watch Co. produced the Auburndale Timer upon finding their rotary watch was not going to be successful. Jason Hopkins, who invented their rotary watch, developed the new movement and a Mr. Craig designed the start/stop mechanism. The watch was wound by opening the back of the case and turning a small winding handle. What appeared to be a winding stem outside the case operated the action of the watch. The timer started when the stem was turned a partial turn in one direction and stopped when the stem was turned back to its original position. Pushing the stem inward returned the sweep hand and recorder hand to zero. There was a sweep hand which rotated once every minute and a second hand at 6 o'clock which rotated once every second. At 12 o'clock was the recorder hand which indicated how many minutes had elapsed, rotating counterclockwise and indicating up to ten minutes. These were made in various models which beat 1/4, 1/8, or 1/10 seconds plus a split second timer having two sweep hands that could be stopped independently. The timer also had a unique escapement adding to its charm and was the most popular watch made by the Auburndale company during its seven years.

159. Movement of Auburndale Timer #10.

160. Manhattan Timer ca. 1885 — hands are set by pulling the button left of the stem, and turning the button. Pulling out the button right of the stem stops the watch.

About the only other timers were made by the Manhattan Watch Company, that sold inexpensive watches in the 1880's with an unusual setting mechanism. The hands were set by pulling out a button beside the stem and then turning the button. A number of Manhattan watches also had sweep hands and some of these were made as timers. On timer models another button, on the other side of the stem, stopped the watch mechanism completely, obviously interrupting timekeeping, so that they were much like the Chronodrometer and apparently no more popular. During World War II Elgin and Waltham also made unusual timers for the government. These had an extremely small balance wheel oscillating at 20 beats per second, giving the sweep hand an almost continuous motion. Depressing the winding stem started and stopped the sweep hand and returned it to zero, but when the sweep hand stopped the mechanism continued running, as with a chronograph.

The design of chronographs, allowing start/stop action while continuing to tell time, had evolved slowly through the 19th Century. Earliest in the chain were double-train timers, which date almost to the beginning of the century and continued to be built into the 1870's. These utilized two mainsprings with separate geartrains, operated off one escapement such that the train used by the timer could be stopped without interfering with the time train. Sometime between 1850 and 1870 the castle-wheel type chronograph was developed, amounting to a transmission with very fine-toothed gears that could shift the timer in and out of mesh with the time train, thus requiring only one mainspring. Though the design was effective, it required careful construction and added many extra levers, springs and screws to the watch. Castle wheel type chronographs were quickly perfected and have continued in use for both chronographs and timers to the present.

Chronographs were first manufactured in the United States by Waltham during the middle 1870's. Initially they sold two different types: a less expensive version on a keywind 16-size movement, and a 14-size stemwind model of high caliber. Both sorts had center sweep chronograph

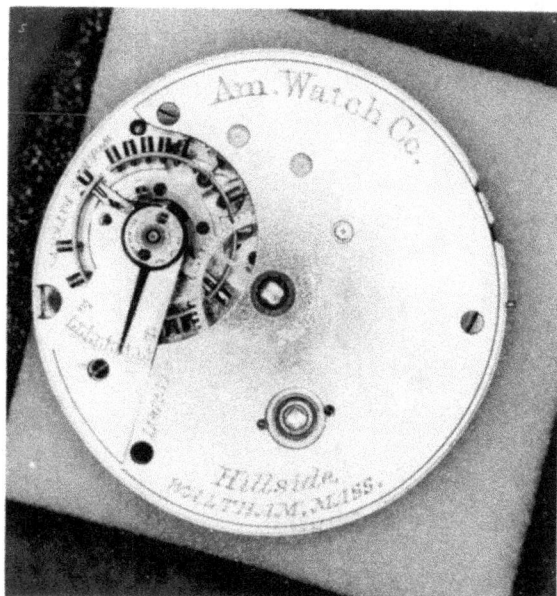

161. Waltham simple chronograph #1,062,511 ca. 1876 — the movement is similar to a model 68. The chronograph, operated by a button at the edge of the case near the pin, starts and stops but does not return to zero.

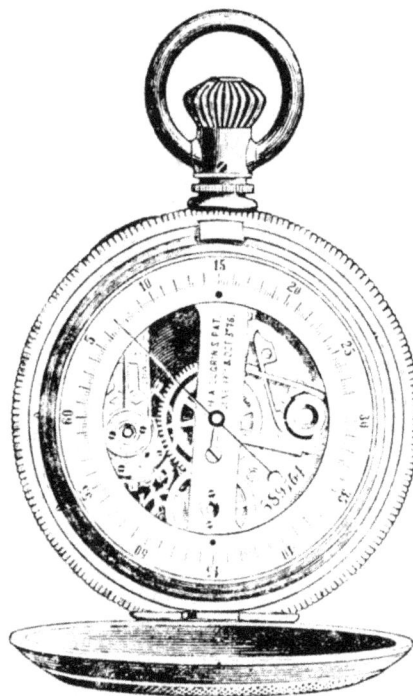

162. Waltham back chronograph ca. 1876 — time dial on the front and chronograph on the back, under a hunting-type lid. Based on Lugrin's patent.

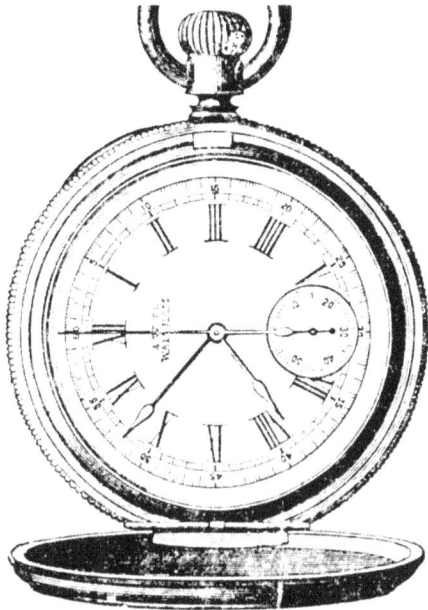

Waltham Chronographs

MADE TO

Start, Stop & Fly Back

hands, plus continuous running hour and minute hands for uninterrupted time-of-day. The cheaper design used a very simple mechanism to operate a sweep hand that would start and stop, but not return to zero. Lacking the "flyback" feature, this was inconvenient to use, yet the price must have been substantial since it was a complication added to a quality jeweled watch. Sales were consequently low and the watch was rapidly discontinued. Waltham's other design, applied to model 74 and 84 movements, was based on a series of patents by H. A. Lugrin, of New York, covering his simplified castle wheel chronograph. These provided start, stop, and flyback actions and obtained a sure meshing of the chronograph train, while having fewer parts than other chronograph mechanisms. A more manufacturable design resulted, which was licensed by Waltham, Timing and Repeating Watch Co. of Geneva, Longines and probably others. Waltham models were the finest of these, commanding $100 in gold-filled cases. First in the Waltham series was the back chronograph, which had a standard time dial on the front of the watch and a chronograph dial on the back. Later models adopted the more conventional arrangement with the sweep hand on the time dial, and some of these had a register which kept track of the elapsed minutes on a subsidiary dial. Waltham chronographs also appeared with split seconds, or two sweep hands which could be stopped independently, and a few with split seconds and split minutes.

Waltham represented the finest chronographs made in this country. Almost as interesting was the Pastor Stop Watch made by the E. Ingraham Co. and sold under the name Sterling Watch Company. This was an inexpensive timekeeper of the dollar watch variety incorporating a full chronograph mechanism with start, stop and flyback ac-

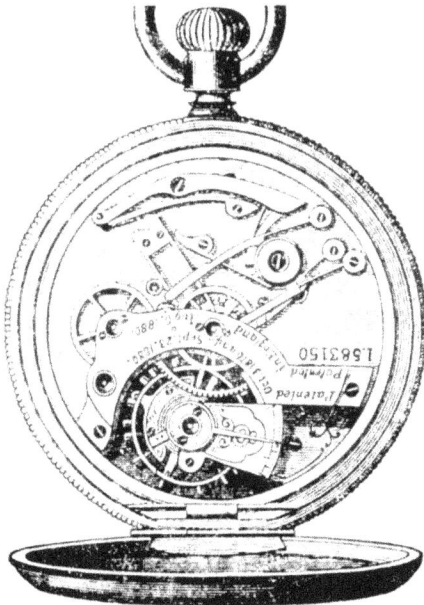

163. Waltham front chronograph based on Lugrin's patent.

164. Dollar watch chronograph made by E. Ingraham ca. 1920 — the dial is marked Pastor Stop Watch, Sterling Watch Co.

99

165a. and 165b. Chronograph movements by New York Standard and New England Watch Co.

also made chronograph watches marked only "Dan Patch Stopwatch" by contract with the horse's owner, Marion Savage, who promoted a number of products under the name of the well known pacing horse.[1]

Manhattan Watch Co. also made sweep second watches with no stop mechanism, known as doctors' watches, simply for the convenience of having a large second hand. There were a few other center second watches in the American industry, both dollar watches and high grade, but the finest were those made by Elgin during the 1880's. These were based on their 16-size movement, appearing in both gilt and nickel finish. Hamilton and Waltham also made some high-quality center second models with hacking feature during World War II.

Another significant area of experimentation was development of anti-magnetic watches. The effects of magnetism had been noticed as far back as 1825 when Jacques Frederic Houriet of Le Locle, Switzerland, observed the derangement of chronometers carried on an expedition to the North Pole. His solution was to use steel only for the mainspring and pinions, with brass and white gold for the compensation balance and hardened 18 karat gold for the hairspring.[2] His efforts received little notice, likely due to the scarcity of polar travel, but later in the century magnetism became an important watchmaking issue. This was because the advent of electricity brought strong magnetic fields produced by generators and electric motors. Magnetism has little effect on today's watches because they use modern alloys developed by Guillaume in France. However steel balance springs of that period were prone to magnetism, and Houriet's approach of using gold for the escapement parts was not practical for large-scale production. The person who solved the problem was Charles-August Paillard (1840-1895) of Switzerland, who was able to identify the right material because of his experience in South America. He had lived with an uncle in Brazil, where he worked as a chronometer repairman. There he observed how salt air and humid climate could rust steel parts, which was a seri-

tions. It was certainly the cheapest such watch available, representing a considerable complication when it is remembered that an inexpensive watch was still expected to be reliable. Through the latter part of the 19th Century a number of other companies came out with relatively inexpensive chronographs, not of the dollar watch type but based on low-cost jeweled movements. The Manhattan Watch Co. marketed a chronograph under the name of the New York Chronograph Watch Company based on their usual 18-size movement. Likewise the Trenton Watch Company, New York Standard Watch Company and New England Watch Company all made chronographs in the moderate price range of $15 to $20. The latter two companies

166. Fredonia non-magnetic #8403 ca. 1883, note worm gear regulator later used by Peoria, successor to Fredonia.

167. Peoria non-magnetic movement, Paillard's patent ca. 1888.

168. Non-Magnetic Watch Co. — Chicago, Paillard's patent — made by Illinois Watch Co. ca. 1900.

ous problem since a tiny bit of rust could ruin the performance of a balance spring. He returned to Switzerland in 1862 and in 1876 began to pursue the idea of rustproof hairsprings, settling on the metal palladium, which was found in certain Brazilian gold ores. By 1877 he was entering commendable timekeepers with palladium balance springs in time trials at the Geneva Observatory, and by the 1880's many such watches were being entered by notable makers. While his intent was to overcome rust, palladium proved immune to magnetism, so that he patented non-magnetic alloys for hairsprings, balances, and levers.[3]

The Fredonia Watch Co. of Fredonia, New York, introduced non-magnetic hairsprings into American watchmaking about 1884. Like a number of companies, Fredonia used imported hairsprings and balances, some of which were non-magnetic, and plates on these watches were marked "anti-magnetic hairspring." While these did not necessarily come from Paillard, it is certainly a good possibility. In 1885 the Fredonia Watch Co. was reorganized as the Peoria Watch Co. of Peoria, Illinois, using Fredonia machinery and making nearly an identical watch. This was one of the companies organized by John C. Adams and through his soliciting had a trade among railroad companies in the days before railroad watch standards. Probably about 1888 some 18-size Peoria watches were made with non-magnetic overcoil hairsprings and marked "Non-Magnetic Watch Co. of America . . . Paillards Patent Balance & Spring." The Non-Magnetic Watch Co. of America was something of a mystery and the Peoria watches were the only American movements to bear this name. In the previous year, 1887, the "Locomotive Engineers' Monthly Journal" had contained an ad for the "Geneva Non-Magnetic Watch Co. . . . Paillard's Patent" showing a Swiss-made, but American-looking, movement fitting an Elgin 16-size case. In 1888 the same movement appeared in jewelery trade catalogues marked "Non-Magnetic Watch Co. of America," along with a second movement even more American in design. The earlier style movement was the higher grade of the two and had the curious quality of meeting many requirements of the railroad standards which were issued about three years later. These included such non-Swiss features as high jewel count, double roller, quick train, overcoil hairspring, and micrometer regulator. Other similarities to American watches also appeared in the movement design. The watches were machine made but apparently hand finished in the finer grades. The "Jewelers Weekly" in 1891 attributed the 16-size watches of the Non-Magnetic Watch Co. of America to Aeby & Co. and J. J. Badollet & Company, both of Switzerland. It can only be said now that such a source seems as likely as any. Meanwhile the Waltham company responded by producing some 18-size model 1883's with non-

magnetic springs and balances plus some 16-size model 1888's.[4] These were higher grade than most movements in those movement styles. Illinois Watch Co. also manufactured non-magnetic watches marked "Paillard . . Non-Magnetic Watch Co. . . Chicago USA." These were offered in a full line of 18-size fullplate and 16-size split plate models. Beyond Waltham, Illinois and Peoria, few American companies made non-magnetic watches.

Some of the most interesting, and certainly uncommon, American watches were repeaters, which rang the time on gongs when activated by a slide on the side of the case. The idea was invented around 1690 by Daniel Quare of London, at that time striking the most recent quarter hour on a bell when a button was pushed. Half-quarter and 5-minute repeaters were developed through the 1700's, worked by pushing down the pendant of the case rather than a button. By 1800 minute repeaters were made and, soon after, the slide mechanism appeared. The first repeaters in American industrial watchmaking were repeating attachments made by Fred Terstegen, to be added to American or foreign movements. He produced these during the 1880's and 90's in his shop in Elizabeth, New Jersey. There were apparently quarter-, five-minute and minute-repeating attachments, and the few known examples have been applied to American factory movements.

Waltham produced five minute repeaters beginning about 1887. They were all built around basic Waltham movements, but the repeating mechanisms were designed and made in Switzerland then assembled to the watch movements at the gold case factory of Robbins and Appleton in New York, sales agents for Waltham. The first repeaters were based on the 16-size model 1872 movement using a repeating mechanism patented by Mr. George Aubert of Switzerland. While Waltham repeaters are known for being five-minute repeaters, a small number of this first series were minute-repeating. A second type was built on the 14-size model 1884 chronograph and housed in an 18-size case. This combined the repeating mechanism of Mr. Aubert with the chronograph patented by H. A. Lugrin, a marriage of mechanisms already being used by Waltham. The third and final form of Waltham repeater was that patented by Charles Meylan of New York and applied to

169. Waltham 5-minute repeater #3,793,994 ca. 1890.

170. Elgin convertible #4,329,466 ca. 1890 — fitting either open or hunting case. The 21-jewel grade of this model was Elgin's highest grade movement.

14-size model 1884 movements. Some of these also had a split second chronograph, not of the Lugrin type but a more conventional Swiss design, and probably constituted the most complicated American watches. The quality of all Waltham repeaters was very high and they were among the most spectacular watches made in the United States.[5]

Another sort of variation came about with the advent of stem winding. Keywind movements could be interchanged from open face to hunting cases with no problem since the location of the winding square was standard, and cases had keyholes in the respective locations. The keyhole location was different for the two cases since the pendant was at 12 o'clock on open face watches and at 3 o'clock on hunters, necessitating the movement being rotated 90 degrees when going from one to the other. When stem winding incorporated the winding crown into the pendant of the case, with associated winding and setting mechanism in the movement, it was no longer simple to interchange movements, since movements would have difficulty accepting stems at both 12 and 3 o'clock. It was not impossible though, and several attempts were made to accomplish this, the nicest being the convertible watches by Elgin. These were the invention of Charles S. Moseley, who invented the hollow split chuck lathe for Dennison and Howard, designed machinery at the Nashua Watch Co. and had gone to Elgin as general superintendent. At Elgin he invented their triangular hairspring stud, dust band, and patent regulator (derived from a regulator conceived at Nashua) as well as the convertible-style movements. The first of these came out in 1880 in a 16-size model and over the years they were produced in 18 size and a variety of grades, including the highest quality watches made by Elgin. To be convertible from hunting to open face, the barrel bridge was designed to cover an arc from 12 o'clock to 3 o'clock and have accommodations for a stem in both locations. When inserted in either, the stem would mesh with the winding mechanism, which also operated the hand-setting device, of the lever-set type. The 18-size movements were really a 16-size backplate with an oversized pillar plate and a heavy brass ring around the movement to make up the difference. The New York Standard Watch Company, which always seemed to be producing something unusual, made a convertible movement along the same line. This was 16 size and in the moderate quality range of all New York Standard models.

Trenton Watch Co. also made a convertible model of moderate quality.

A less complete approach was taken by the Illinois and Aurora watch companies. They designed movements with only one stem and a geartrain laid out for a hunting movement. There were also pivot holes in the plates for an extra pinion which was located to carry the second hand for an open face watch, idling off the third wheel of the train. When this extra pinion was used, the arbor of the fourth wheel was left short since it did not have to carry the second hand, and a different dial was used with the seconds bit located appropriately relative to the dial feet. Thus the difference between open face and hunting movements was minimal and the company could use common parts. The purpose of convertible and extra pinion movements was to boost wholesale sales by offering jewelers flexibility while keeping fewer movements in inventory, therefore lowering total inventory cost. Unfortunately, the greater complexity

171. New York Standard convertible model ca. 1890.

172. Dudley Masonic watch — modern movement finished by X-L Watch Co. of New York from recovered Dudley material.

102

meant such movements were higher priced, which Elgin convertibles certainly were. In the end, all such ideas faded with the decline of hunting watches and companies succumbed to the inevitability of producing completely separate movements for hunting and open cases.

Another unusual group was the rotary watches made by the Auburndale and Waterbury watch companies. The first rotary watch was reputed to have been built by George P. Reed in 1862 and called the Monitor.[6] Where he got the idea was unknown but he certainly was an independent and excellent watchmaker, capable of arriving at the design himself. Auburndale's rotary was designed by Jason Hopkins of Washington, DC, and the Waterbury rotary was the invention of D. A. Buck of Worcester, MA, who invented the American-style inexpensive duplex escapement to go with it. Since these mechanisms rotated inside the case, they were often compared to tourbillon watches invented by Breguet in Paris. Tourbillons averaged position errors by mounting the escapement in a small carriage which revolved once each minute. This was an expensive and delicate device used to achieve precision timekeeping, so that many people asked why it was used in low-grade watches such as the Waterbury and Auburndale. In both instances the intent was to make a simple timekeeper with fewer parts than ordinary watches, and therefore cheaper. Whereas the conventional stemwind watch had 150 or more parts, the Waterbury got down to 54. It could be argued that the Waterbury, with a crude stamped balance, used the tourbillon action to average the large position errors such a balance was bound to have. Waterbury overcame or ignored any such problems and made non-rotating watches also. In any event inherent economy, rather than timekeeping, made the American rotary watch popular. Early Waterbury watches were unusual in being skeletonized for an interesting appearance. Few other American watches had the skeleton look except the New England duplex. While millions were made, the Waterbury watches were so cheap that most were thrown away over the years, and they've nearly all disappeared.

The Masonic watches of William Wallace Dudley were among the unusual articles in American watchmaking. Dudley (1851-1938) came to America after apprenticeship to a chronometer maker in his native New Brunswick, Canada. Here, he worked successively for Waltham, Illinois, South Bend, Trenton, and Hamilton. The Dudley Watch Company began in 1920 to make pocket watches having bridges in the form of Masonic symbols. Winding/setting mechanisms and bridges were made in Dudley's Lancaster, PA, plant, but other parts were purchased from various foreign and domestic sources. First watches were based on a 14-size Waltham model 97 movement, then a smaller 12 size was modeled from Hamilton model 910, 912, and 914 material. At a base price of $125, Dudley Masonic watches were expensive at a time when wristwatches were eroding demand for pocket timepieces. So, sufficient support to sustain such a watch factory was not to be found among Masons. The company went into receivership in 1925 and the new owners attempted importing a Swiss wristwatch to survive. In 1929 the foundering concern was bought by a jewelery manufacturer, who disposed of old Dudley inventory at low prices during the Depression. The watch machinery and parts inventory were then sold in 1935 to Mr. Menche of the X-L Watch Company of New York. Unable to unload the machinery, the X-L Watch Company finished 12-size Masonic movements till around 1970, completing about 1000 units. Combined with approximately 2600 made by Dudley, total production of Masonic watches came to about 3800 over a 50-year career.[7]

Among the most aesthetic of oddball watches manufactured in American factories were the crystal plate movements made at Waltham. Advertised as stone movements, these had natural crystal back plates set over gilded, damaskeened pillar plates. With their transparent backs, the gold train and ruby jewels seemingly floated above the ornate bottom plate. These were made during the late 1880's and early 1890's by William R. Wills, foreman of the jewel department. Mister Wills, naturally interested in minerals and lapidary, purportedly returned from a Mexico vacation with a considerable supply of agate and rock crystal.[8] From these he fashioned possibly as many as 200 stone movements, mostly in 6 size with crystal back plates only. Waltham catalogues of the early '90's advertised stone-movement model 72's with either crystal backs, or crystal backs and fronts. "The American Jeweler" magazine reported in brief notes during 1888 that several interesting stone movements were in process.[9] One was a model 72 having a crystal back with gold damaskeened pillar plate and winding wheels. Based on that success came a model 72 having a crystal back with ruby jewels, a striped agate pillar plate with sapphire jewels, non-magnetic balance, skeleton dial having rubies at the minutes and sapphires at the hours, and an 18K display case. This feat was followed by another sporting crystal front and back plates, ruby jewels in settings of sapphire, skeleton dial having rubies at minutes and diamonds at hours, and a gold display case. A diminutive version was a 4 size with crystal back plate, set in a display case carved from one piece of crystal. While expensive, these were not practical, frequently suffering from breakage. As many as a dozen remained in the factory during the 1930's, and the final one

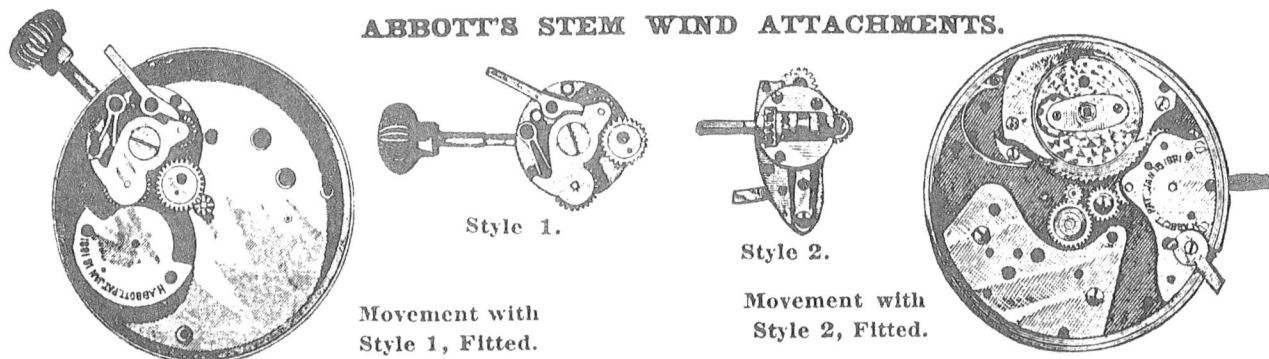

ABBOTT'S STEM WIND ATTACHMENTS.

Style 1.

Movement with
Style 1, Fitted.

Style 2.

Movement with
Style 2, Fitted.

173. Advertisement for Abbott stem wind attachment ca. 1885 — style 1 is a rocking bar mechanism, style 2 is a shifting sleeve mechanism.

was sold for $10 in 1939, the balance cock having been broken and reglued.[10]

One of the remaining oddities receiving occasional attention was the "butterfly" cutout in early watches by the United States Watch Co. of Marion, New Jersey. Efforts by the Marion company to produce a good quality watch resulted in a number of fine timepieces, and an important area of attention was adjusting of escapements. This must especially have been a problem in that early period when manufacture of duplicate parts was not yet perfected and each set of escapement parts might have required individual adjustments. Since the escapement was difficult to see in a fullplate watch Mr. Frederick A. Giles patented the butterfly cutout to allow observation of the escapement operation after the watch was assembled. It was thus functional as well as decorative. Actually this was not the first effort to aid escapement adjusting. As Dennison and Howard were building their first watches in Roxbury, Nelson Stratton introduced two peep holes in the pillar plate, in line with the pallet jewels, so the escapement action could be verified by the adjuster. These were used by Waltham for years afterward as well as by Elgin and other companies.

While the mechanisms covered above were obviously different, there were numerous devices of a more subtle sort. The Rockford Watch Co. built stemwind watches which could become either stemset or leverset by making an adjustment on the back of the movement. Also there were the more general categories such as transition period movements which were stemwind but had a key winding square, and transition cases which had a stemwinding pendant but also a keyhole in the back. Other general areas were the almost limitless types of regulators used, plus a maze of patented mainspring barrels and saftey pinions. Likewise winding/setting mechanisms were subject to wide variation due to numerous patents covering them. For instance early pieces by the New York Standard Watch Co. employed a giant winding wheel in the form of a ring with internal teeth placed under the dial and the diameter of the entire movement.[11] Also along the line of winding mechanisms was Abbott's winding attachment, patented in 1881. This was not the product of a watchmaking company but was made by a watch dealer located on Maiden Lane, New York. Henry Abbott made a rocking bar winding attachment which could be added to keywind watches which were still common at the time.[12] This created a stemwind watch and functioned as a lever set device so that the hands could be set by the stem as well. The movement could then be placed in a stemwind case to continue life as a modern new timepiece. Beyond all this were variations in escapements, which will be covered in the next chapter.

Many oddball watches were simply wrong turns in the development of American watchmaking, improvements which didn't turn out to be improvements, so went rapidly by the wayside. Some were novelties brought out by new companies to attract attention, or innovations from established companies to attract new business. Others were temporary measures, such as transition period movements, to accommodate the trade while changes were taking place. Such circumstances promoted manufacture of the unusual, providing interesting variations in an industry which otherwise drifted into boring repetition.

REFERENCES

1. Warner D. Bundens, *The Story of the Dan Patch Watch* (BULLETIN, National Association of Watch and Clock Collectors, Inc., April 1975), p. 181.

2. Jaquet & Chapuis, *Technique and History of the Swiss Watch* (Olten, Switzerland, 1953), p. 202.

3. Ibid., p. 203.

4. Thomas DeFazio, *A Note Concerning the Non-Magnetic Watch Company of America* (BULLETIN, National Association of Watch and Clock Collectors, Inc., February 1975), p. 39.

5. Gerrit A. Nijssen, *The American Repeater* (BULLETIN, National Association of Watch and Clock Collectors, Inc., December 1973), p. 4.

6. Charles S. Crossman, *The Complete History of Watchmaking in America* (reprint, Exeter, New Hampshire, Adams Brown Co.), p. 193.

7. Stoltz & Parkhurst, *William Wallace Dudley and His "Masonic Watch"* (BULLETIN, National Association of Watch and Clock Collectors, Inc., October 1968), p. 496.

8. Personal correspondence of J. E. Coleman to E. T. Fuller, April 1970.

9. J. E. Coleman, *Foreman Wills' Crystal Watches* (BULLETIN, National Association of Watch and Clock Collectors, December 1954), p. 311.

10. A. E. Mathews, *A Crystal Plate Waltham Watch* (BULLETIN, National Association of Watch and Clock Collectors, Inc., April 1971), p. 1075.

11. Paul M. Chamberlain, *It's About Time* (New York, Richard R. Smith, 1941), p. 104.

12. Elsworth H. Goldsmith, *Keyless Watches* (BULLETIN, National Association of Watch and Clock Collectors, October 1953), p. 496.

174. New York Standard Watch Co. worm drive #28,155
ca. 1887 — worm is visible through star-shaped cut-out in
back plate.

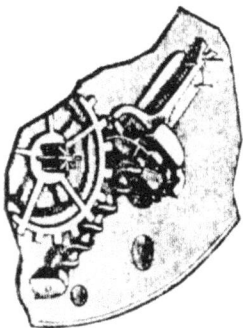

Escapement Enlarged.

SEND TO US FOR SAMPLE
WATCH WITH A
WORM IN IT
WONDERFUL
NOVELTY
BEAUTIFUL IN
DESIGN
HONESTLY AND ELEGANTLY MADE
ACCURATE TIMEKEEPER
NEW YORK STANDARD WATCH CO
— 83 NASSAU ST., NEW YORK —

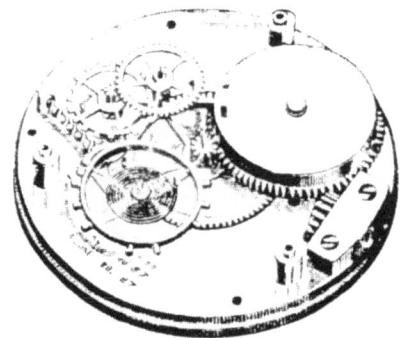

Movement with Top Plate Off.

175. Advertisement for New York Standard worm drive, showing train layout and upright escape wheel with worm
rather than pinion.

IX
ESCAPEMENTS

The escapement action, including the balance wheel and hairspring, is the heart of any watch since the stability of timekeeping is generated there. Two main tasks must be performed by the escapement: 1) provide impulses of energy to the balance to keep it oscillating, using power transmitted through the geartrain from the mainspring; and 2) allow oscillations of the balance to let the train run down at a constant rate, one tooth at a time. The 18th Century had been an age of escapement invention, following closely behind discovery of hairsprings, when numerous mechanisms were devised for accomplishing these tasks. In the 19th Century the best of these were refined for better performance and easier manufacture, and during the early years of the American watch industry some of this perfecting process was still going on. This was not easy going since the escapement entailed some of the most delicate parts of the watch, while at the same time the most complex, with violent action, impact and wear. One of the objectives was to reduce friction, since there is no use storing energy in the mainspring only to waste it scraping away the escapement. Another was to perfect designs that permitted the balance to oscillate in an ideal manner for good timekeeping. For their finest timepieces the English used spring detent escapements while duplexes and levers served for moderate to high grade watches. The Swiss were making levers and pivoted detents in their best watches, with cylinder and lever escapements for low and moderate grade pieces. Levers have thus been nearly universal in American factory watches since English and Swiss watchmakers had already developed them to be efficient and producible. There have been a number of variations on the lever as well as some completely different escapements tried by the factories, in addition to escapements used by individual makers who were not limited by industrial methods. As with other types of variations, these escapements are not often found and are always interesting when they do appear.

In 1810 Luther Goddard, who made the first well documented watches in America, used the verge escapement, oldest known escapement in watch or clockmaking. The verge was very sensitive to variations in motive force, therefore such watches always utilized a fusee on their mainspring to keep driving force constant as the mainspring ran down. The verge was also the simplest escapement to produce, which was probably why Goddard used it, and why it persisted in a few foreign watches as late as 1880. Unfortunately, it was also among the poorest of timekeepers. Watchmakers long ago discovered that the balance wheel and hairspring, or balance spring, were central to precision timekeeping. Once properly adjusted these wanted to remain as undisturbed as possible to maintain a constant rate. Verge escapements severely encumbered the balance wheel, but after invention of the hairspring in 1675 it was possible to introduce escapements which left the balance much more free. First came frictional rest escapements such as the Debaufre, cylinder, virgule and duplex. These were all like the verge in locking the escape wheel against the moving balance staff, which intermittently allowed one escape tooth to pass as the balance oscillated. Admittedly, they did this in much more subtle fashion than the verge, reducing interference with balance action and allowing more lively motion. Still they imposed a low level of constant frictional drag on the balance, hence their name, frictional rest escapements. Balance wheels weren't fully free till invention of detached escapements. The lever and detent (or chronometer) were the two basic detached escapements, all others being variations on these. Being detached implied that an additional piece was introduced for locking the escape wheel, either a lever or detent. This relieved all frictional drag from the balance, leaving it only the interference of unlocking the escapement and receiving impulse. Thus liberated, balances could oscillate more ideally for proper timekeeping, with temperature compensation and hairspring adjustment to eliminate other detracting influences. These inventions were combined to produce precision pocket chronometers by 1800.

While detent escapements were being used in precision pocket timepieces, there was limited demand for such watches and the lever escapement emerged best for general timekeeping needs. Levers provided accurate timekeeping yet could be produced at reasonable cost. The detached lever was invented in 1754 by Thomas Mudge of London, an exceptionally skillful watchmaker, based on the rack lever escapement which was not detached at all. Unfortunately his design was so unnecessarily complicated that even he avoided using it, and little came of detached levers for another 50 years. In 1800 Breguet of Paris was refining his version of the lever, which resulted in its popularity among Swiss makers by 1840. As of 1800 the English had regressed to the rack lever, not embracing detached levers till 1815 when the Massey and Savage 2-pin designs were patented. English lever escapements were quickly perfected and by 1850 detached levers were markedly prominent in all European watchmaking. These designs were economical to produce since they lent themselves to subdivision of labor in the European handcraft system. Likewise the individual parts were simple enough to be made by factory methods, so lever escapements were popular in the American industry. At the same time, levers could be made well enough to meet modern timekeeping requirements.

The earliest American lever escapements were in watches by the Pitkin brothers of East Hartford during the late 1830's. Beside being the first lever watches, they used pointed tooth escape wheels, which rarely appeared in American watchmaking.[1] Pointed teeth were prone to damage, even before leaving the factory. Thus while Dennison and Howard started with pointed tooth-escapements, Nelson Stratton, who had worked for the Pitkins, convinced them to switch to club-tooth escape wheels such as Swiss watches used. Even when the Pitkins were making their watches, Swiss imports existed which would have made the idea apparent, but English watches were the finest available and since English makers used pointed escape teeth, so did the Pitkins. Edward Howard used pointed teeth in some early watches and so did Belding Dart Bingham in the 1850's, before he joined with Nelson Stratton in the Nashua Watch Co. since his design was based on London-made watches.[2] Likewise Ezra Bowman used them for watches he made in Lancaster, PA, around 1880 since his escapement was modeled after fine English levers by Frodsham. A bit later the Manhattan Watch Co. used pointed wheels in their watches, but this was more likely a holdover from escape wheels used in the clock industry, since the chief management of that company had previously been with the Jerome Clock Company.

The form of lever escapement which evolved in American watches developed from both English and Swiss designs. The English used exclusively side levers with covered pallets, while the Swiss used both side and straight-line levers with both covered and exposed pallets. There were other differences between the two countries, for instance the English used only pointed-tooth escape wheels while the Swiss used only the club-tooth form. While trying almost everything at one time or another, American factories settled on a fairly Swiss-looking design with a club-tooth escape wheel and straight-line lever. Side levers were often used for fullplate movements and American balance wheels were

Fig. 26

LEVER ESCAPEMENTS

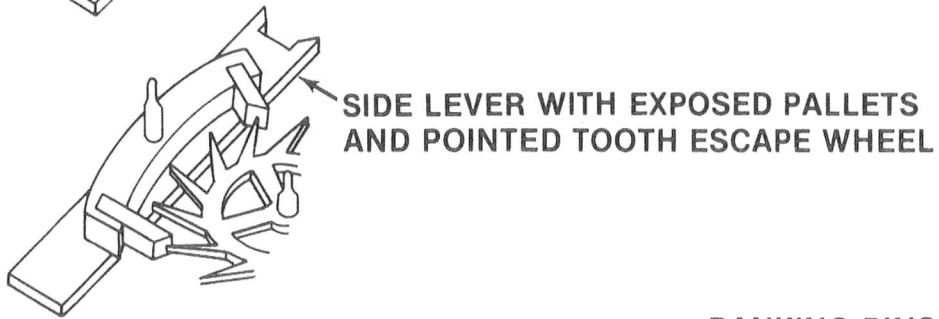

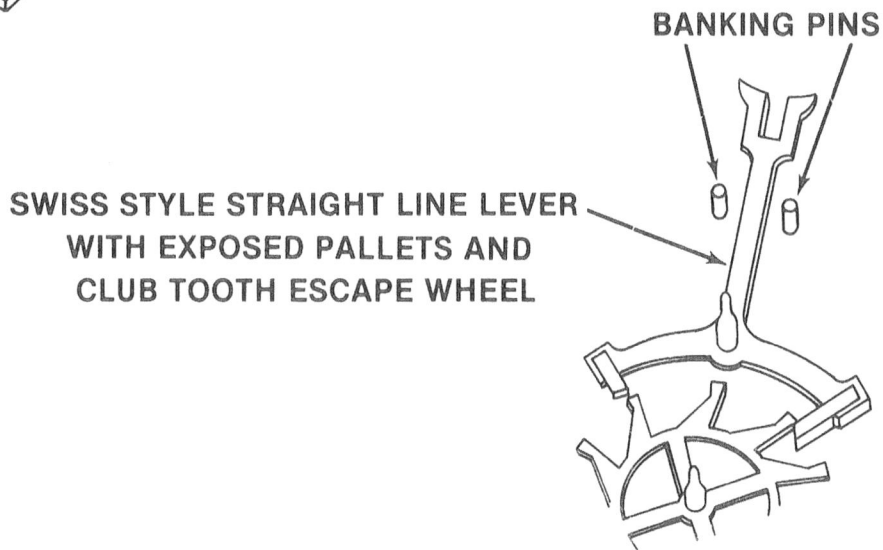

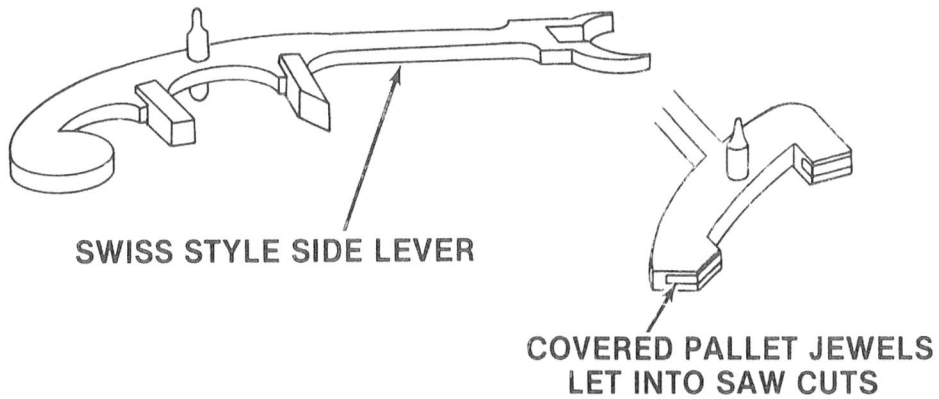

ENGLISH STYLE SIDE LEVER
WITH COVERED PALLETS

PALLET JEWELS LET INTO SAW CUTS

SIDE LEVER WITH EXPOSED PALLETS
AND POINTED TOOTH ESCAPE WHEEL

BANKING PINS

SWISS STYLE STRAIGHT LINE LEVER
WITH EXPOSED PALLETS AND
CLUB TOOTH ESCAPE WHEEL

SWISS STYLE SIDE LEVER

COVERED PALLET JEWELS
LET INTO SAW CUTS

Fig. 27

**LONG LEVER
¾ PLATE LAYOUT**

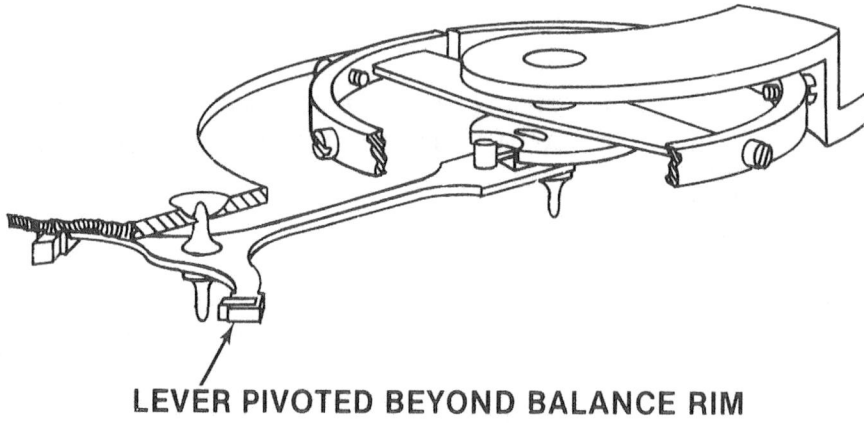

LEVER PIVOTED BEYOND BALANCE RIM

Fig. 28

SHORT LEVER

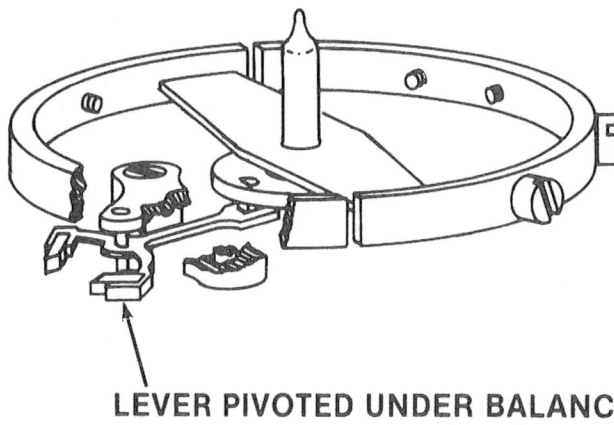

LEVER PIVOTED UNDER BALANCE WHEEL

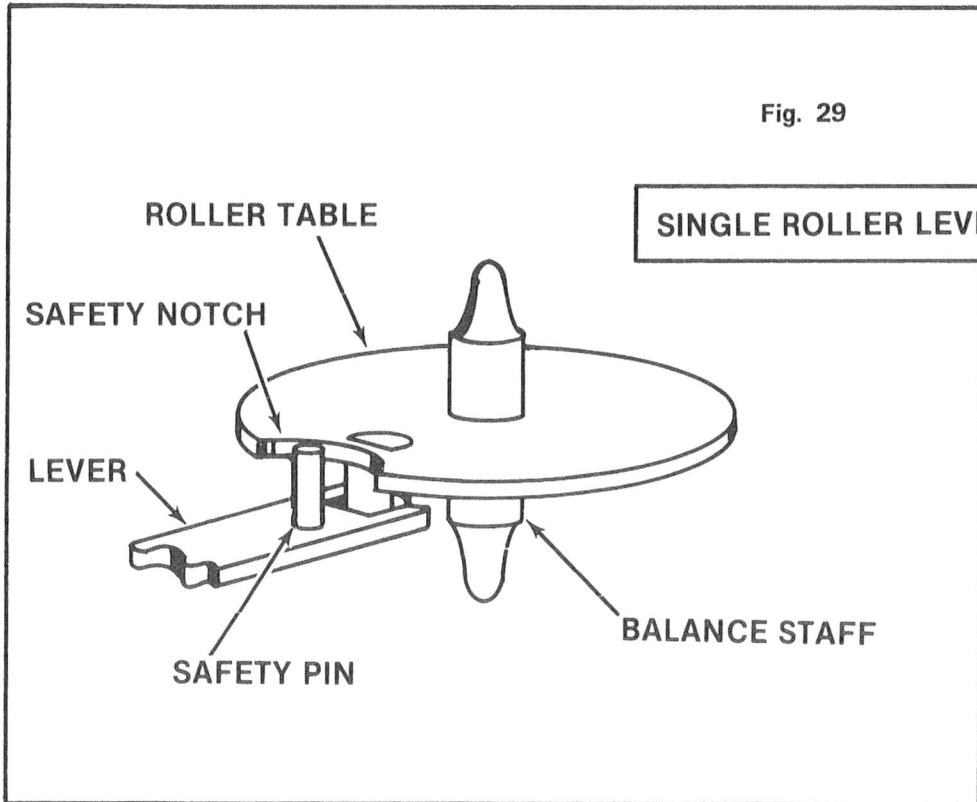

Fig. 29

ROLLER TABLE

SINGLE ROLLER LEVER

SAFETY NOTCH

LEVER

SAFETY PIN

BALANCE STAFF

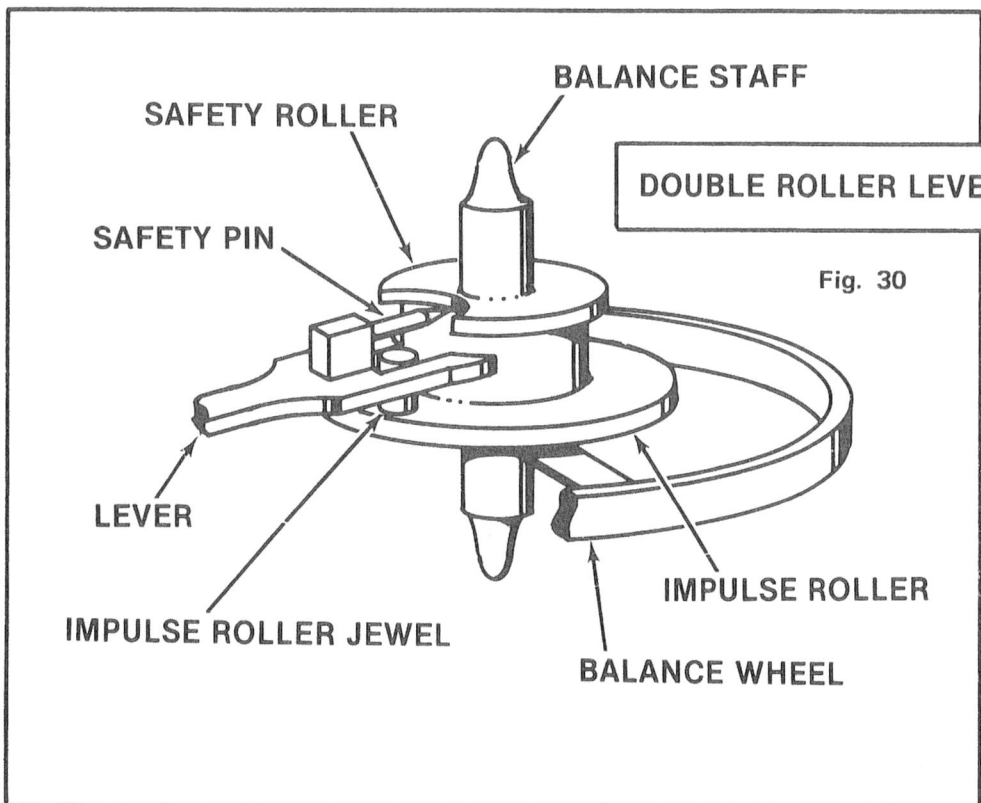

BALANCE STAFF

SAFETY ROLLER

DOUBLE ROLLER LEVER

SAFETY PIN

Fig. 30

LEVER

IMPULSE ROLLER JEWEL

IMPULSE ROLLER

BALANCE WHEEL

176. Waltham vibrating hairspring on KW16 #125,407 ca. 1864 — the vibrating hairspring was applied to Appleton Tracy & Co. grade and American Watch Co. grade movements. The latter is illustrated, having a jewel in the balance cock for the vibrating arm pivot. The vibrating arm extends along the balance cock to the hairspring.

smaller than the usual Swiss practice. American factories also stuck mostly to short levers, in which the lever was tucked down under the balance wheel, even though some high grade Swiss watches used a long lever pivoted beyond the rim of the balance. There were a few attempts to use long levers, such as early ¾-plate movements at Waltham, but these were short lived. Longwinded discussions debated which was more desirable and there was probably no correct answer. Both obtained excellent performance in well-made watches, although American factories did not seem to have a touch with long levers, which were more delicate to produce. Endless debate also raged over the choice of single-roller vs. double-roller escapement. The balance had to carry a roller jewel to interact with the lever, in addition to having a safety mechanism that prevented the lever from being jarred out of position. Both functions could be incorporated into a single roller on the balance staff, or a different roller table used for each, i.e. a double roller. The English almost always used a single roller and so did the Swiss. American companies followed along, avoiding double rollers till they were written into railroad watch standards in the 1890's. It is difficult to determine the origin of this mandate, but the result was to institutionalize double rollers as one of the features in good American watches. As with long versus short levers, both single and double rollers have been used with fine results. In the end, more rested with quality of work than the specific mechanism and American factories built excellent watches of both types.

One of the earliest escapement variations was also one of the most unique, used in watches Charles Fasoldt began making about 1850. Fasoldt called this a chronometer but it was really a form of lever. Recalling the two chief tasks of the escapement, Fasoldt's design had two escape wheels, one above the other on the same arbor. The top one delivered impulse while the bottom wheel was solely for locking and unlocking the train. The chief advantage was that it delivered impulse with less friction than conventional levers, but like a duplex had the problem that it gave impulse only on every other swing of the balance. Fasoldt's escapement, with two escape wheels and rather heavy construc-

tion, was also sluggish in action, so a large mainspring was required to drive the watch. Even so, the fine quality of Fasoldt's work resulted in good performance from his watches as well as a number of clocks he built with this escapement.

Edward Howard was given to experimenting with all sorts of ideas which he thought might produce a better watch, and was more inclined to introduce hopeful improvements than increase production. At one time or another he tried helical hairsprings, upright pallets, Mershon's regulator, and a number of methods for eliminating the usual two banking pins. His singular aversion for banking pins is difficult to explain, although he may have felt they contributed to overbanking. One means of eliminating them was to use a single pin hanging from the pallet frame and banking against the sides of a hole in the pillar plate. Most likely this idea was borrowed from Adolph Lange of Germany, an excellent watchmaker, for Howard also tried other Lange tricks such as concave and convex impulse faces on the pallets to achieve equal impulse from both jewels. (Lange in turn had gotten the latter idea from Breguet of Paris.) In some other movements Howard banked the back of the pallet frame against the pallet bridge, which would have required careful tuning of the bridge for each watch to match that particular escapement. One of his more unusual digressions, which had a side effect of eliminating banking pins, was use of Cole's resilient escapement. This had been invented about 1830 by a superlative English watchmaker named James Ferguson Cole, and a similar but less robust version was used years later by the English firm of Nicole Neilson. The purpose of the escapement was to prevent overbanking, in which the balance lodged itself behind the back side of the lever, causing the watch to stop. This occurred when the lever managed to snap over to the wrong banking pin, implying some maladjustment of the safety action in the escapement. Sudden shocks were the other major factor in causing overbanking, which was undoubtedly the main culprit when the ordinary form of transportation was bounding about on the

177. Waltham vibrating hairspring on KW20 #50,018 ca. 1862 — perhaps an early prototype. Since a standard regulator with curb pins is used, the vibrating arm would not be able to allow free oscillation of the hairspring studding.

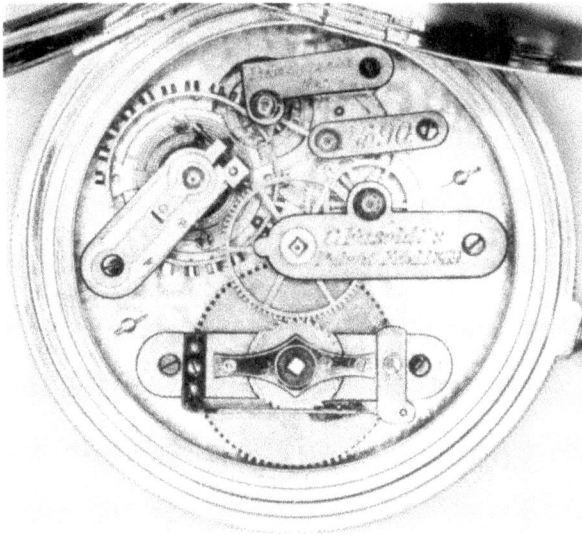

178. Charles Fasoldt #90 ca. 1864 — key wind with Fasoldt's form of regulator and lever escapement. This is an early appearance for a micrometer regulator.

back of a horse. It can be seen why Mr. Howard considered the resilient escapement a useful item. In Cole's design the lever banked against the escape wheel teeth, rather than against firmly planted pins. If overbanking occurred, the force of the balance hitting the side of the lever would cause the lever to deflect out of its way by recoiling the escape wheel backward. In other words, were there not a hairspring the balance could turn continuously in one direction, deflecting the lever out of the way every time it passed.[3] Howard first used this escapement in N size key-wind watches after 1865 and tried it again in some L size stemwinding watches about 1870. Production was never large and many of the keywinders are found to have banking pins added. Later stem models were rejected by the New York group of wholesalers who acted as Howard's sales agent during the period, on the grounds that they were different and would not be received well in the trade. Many were consequently returned to the factory and converted to conventional escapements, some of which performed poorly. These had the Howard name machined off the plates, were renamed "Prescott" after the street on which the factory was located, and sold in the trade at a loss. (Prescott St. was named after Colonel William Prescott, who commanded American troops at Bunker Hill.) Howard watches were about the only production of Cole's resilient escapement anywhere, though a few pieces were made in Europe and Waltham serial number 12,000,000 also used it.[4]

It was mentioned that upright pallets appeared in some Howard models. Just what advantage was expected is not clear, but they were used in some other American watches as well. The next example was most likely by the Manhattan Watch Co. which was formed in 1883, making use of a pointed-tooth escape wheel and upright pallets in the form of semicircular steel pins. The Manhattan company was founded by Mr. A. O. Jennings, who had previously been manager of the Jerome Clock Company, and it will be recalled that the clock industry made wide use of pointed-tooth escapements. Clock companies also used the Brocot escapement, usually placed conspicuously near the middle of the dial. This escapement had pallets constructed of large upright semicircular jewels, and may have been the

inspiration for the Manhattan escapement on the idea that it would be easy to make. This was apparently the case since Manhattan used it for a number of years, presumably adjusting escapements by setting steel pallets to the desired positions. This was certainly better luck than was experienced by Charles DeLong of the Illinois Watch Company. As master watchmaker at Illinois he had a hundred watches made by that firm during the first quarter of this century, using his escapement with upright pallets. DeLong found that semicircular pallets had a theoretical advantage of obtaining effectively large draw with little recoil of the escape wheel. Draw angle kept the lever held against its banking pins so that it could not shake loose, but this was accomplished at the expense of recoil. This meant that as the escapement unlocked, the escape wheel was forced to recoil backward against the mainspring. By reducing recoil DeLong's escapement had higher mechanical efficiency, and a brisk balance motion was obtained with a $3/4$-strength mainspring.[5] In addition, the weaker mainspring reduced forces throughout the geartrain, which also reduced wear. These escapements were interchangeable with conventional lever escapements in Abe Lincoln model watches where they were tried, so that when they failed to reach production some of the 100 may have been switched. Technical problems were reported as the reason for rejection, but the greatest difficulty was most likely the fact that DeLong escapements were different.

One of the greatest gimmick salesmen of American watchmaking had to be Robert J. Clay of Jersey City. In 1883 he was co-founder of the Williamstown Watch Co. of Williamstown, Massachusetts, holding patents for its unique worm drive escapement, while a Mr. F. P. Markham acted as capitalist. Sufficient capital was not to be found till a more conventional 16-size movement was proposed, and the ill-conceived company then muddled to its demise without ever producing a watch. But Mr. Clay's escapement made its public appearance in 1887 with the New York Standard Watch Company, creating the watch with a worm in it. Instead of a conventional pinion, the escape wheel had a polished screw driven by the fourth wheel, and the escape wheel was in a vertical plane. This drove a somewhat conventional lever escapement so that the net result was only to add complication to the normal operation of a watch.[6] There may have been 15 to 30 thousand of these made before the design was discontinued, but the New York Standard watches were an inexpensive grade and many were simply thrown away when they quit running, which would not have taken long. The wonderful worm was readily visible from the back of the movement through a star-shaped cutout in the backplate. The company also used an unconventional winding mechanism as well as later designing a movement which was convertible from open face to hunting cases, like convertible model Elgins. In the long run, New York Standard reverted to a fairly normal watch to make a dependable living till going out of business in 1929.

Another escapement which never found its way to success was the chrono-lever of Don Mozart. In 1864 he formed a company in Providence, Rhode Island, to make watches of his own design. When financial backers noted continued lack of progress, he was removed, and the company continued without him to become the New York Watch Co. and then Hampden. Undaunted, Mozart went on to Ann Arbor, Michigan, and began another company in 1868 with fresh financing. The Ann Arbor company lasted till 1870 when his backers again decided to pull out. Perhaps 30 movements were made in Ann Arbor, a few of which were presented in gold-filled cases to investors. One unusual feature with these was a three-wheeled train, perhaps inspired by

STANDARD LEVER

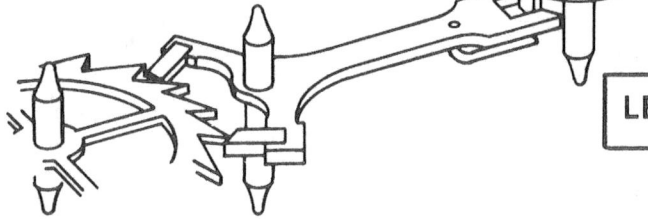

Fig. 31

LEVER ESCAPEMENTS

UPRIGHT
SEMICIRCULAR PALLETS

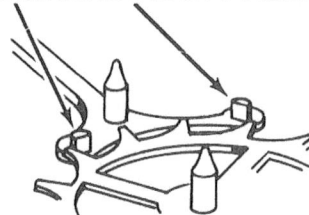

DELONG LEVER

IMPULSE TEETH

LOCKING TEETH

FASOLDT LEVER

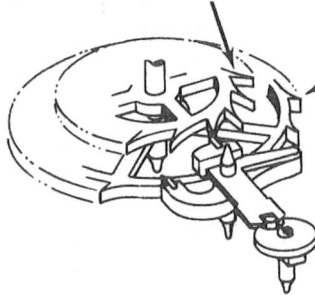

PIN-PALLET LEVER

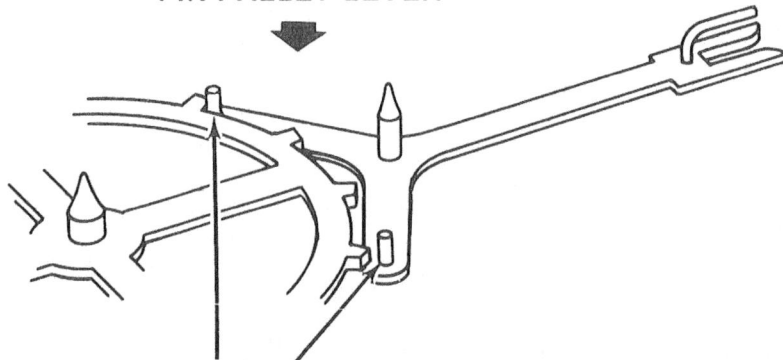

PALLET MADE FROM VERTICAL STEEL PIN

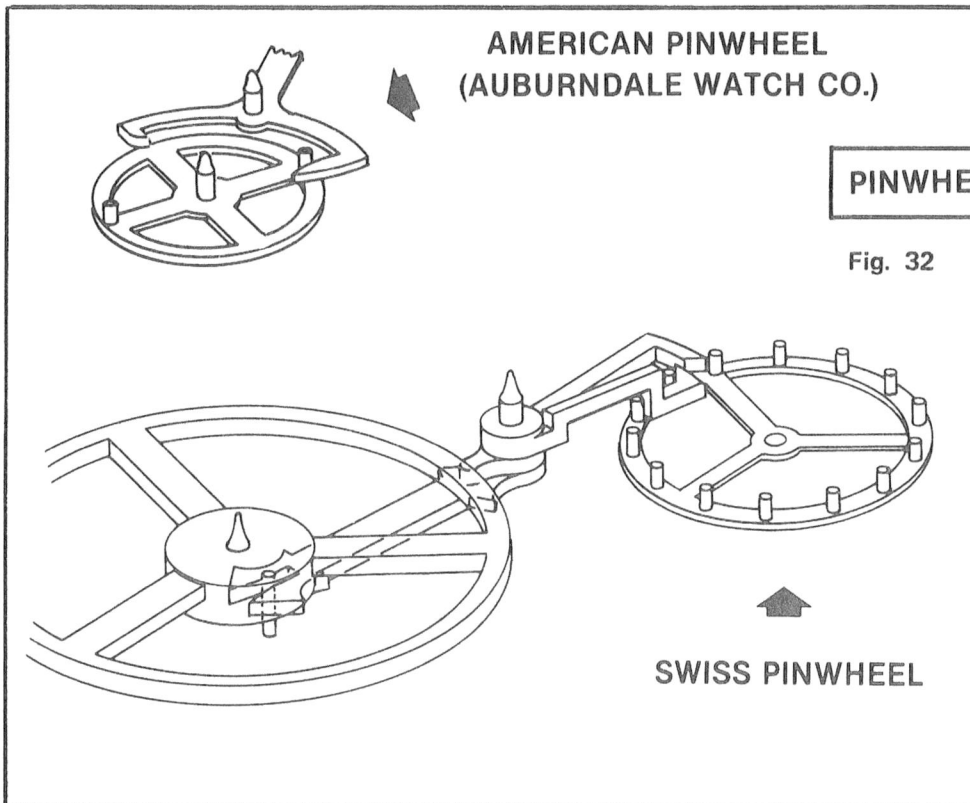

AMERICAN PINWHEEL
(AUBURNDALE WATCH CO.)

PINWHEEL

Fig. 32

SWISS PINWHEEL

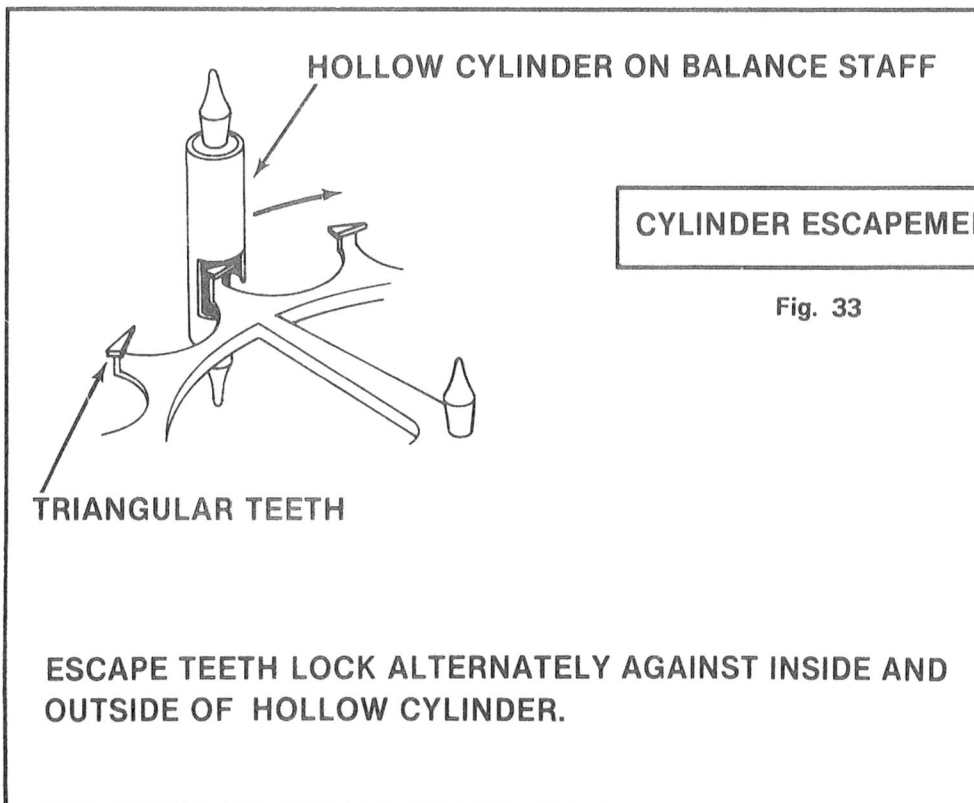

HOLLOW CYLINDER ON BALANCE STAFF

CYLINDER ESCAPEMENT

Fig. 33

TRIANGULAR TEETH

ESCAPE TEETH LOCK ALTERNATELY AGAINST INSIDE AND
OUTSIDE OF HOLLOW CYLINDER.

similar watches from England at the beginning of the century, with a second hand on the escape wheel arbor rotating once every 12 seconds. But more unusual was the escapement. Its main function was to deliver impulse directly to the balance from the escape wheel, as with a chronometer escapement, but overcome the chronometer's shortcoming of delivering impulse only on alternate beats. To accomplish this Mozart used a lever for locking the escape wheel. Impulse was delivered directly by the escape wheel when the balance oscillated in one direction, and through the lever when the balance oscillated back the other way. While he apparently thought the three-wheeled train would save some money, he more than consumed the savings in complex jeweling of the escapement, which would have been difficult to make and almost impossible for the average jeweler to repair. Certainly the escapement was completely original with him, but was better suited to an individual watchmaker than the factory setting in which Mozart tried to apply it.

The pin pallet escapement used in dollar watches could be considered a form of upright pallets, and since the pins were round it secured the advantages set forth by DeLong, namely greater locking with less recoil. Pin pallets were originally devised by Louis Perron (1779-1836) in 1798 with the intent of making a quality escapement, but it was displaced by the cylinder in continental watchmaking. George Roskopf then used the pin pallet for his inexpensive watches in 1868 and obtained good performance. The weakness in the design was difficulty in keeping oil on the escapement since it tended to run away down the pins, but despite this, thousands of pin pallet watches are still made today that run nicely.[7] The chief advantage was that it lent itself to factory production at low expense and could run tolerably even when poorly made. More unusual than pin levers were pin wheels, in which escape wheel teeth were upright pins in the rim of the wheel. This was most often encountered in tower clocks and Swiss regulator clocks found in jewelry stores, using the form known as Amant's escapement. The design was rarely used in watches, where the accuracy of locating numerous pins was a problem on such small escape wheels. A clever exception to this was the Auburndale Timer, made from 1876 to 1883, using a simplified pin wheel escapement. This version was invented by James Gerry, a mechanic and manager for many watch companies, using either two or four pins in the escape wheel and an unjeweled lever.

The mechanisms discussed thus far have all been variations of the lever escapement, which is by far the most widely produced in the world. By 1850 the lever had established itself in Europe, depressing use of the cylinder, duplex, and chronometer, and was being adopted by the rising American industry. Its only challenge since that time was the brief but spectacular rise of the inexpensive American duplex escapement, invented by D. A. Buck for the Waterbury Watch Company. Subsequent use of any other escapement, in Europe or America, has almost always been the effort of an individual maker. Other escapements in American watchmaking thus fall outside the factory system and are the products of lone craftsmen. Generally they are along the lines of traditional escapements; cylinder, duplex, and chronometer.

The cylinder escapement can be easily covered since there have been exceedingly few made in this country. It was invented in 1725 by George Graham of London, who had been the prize pupil of Thomas Tompion and eventually took over the Tompion business. Through the next century it was used in English watches of better-than-average quality, but quantity was always limited because the escape wheel was difficult to make and the cylinder suffered wear.

The firm of Breguet in Paris, for whom difficulty and expense were not s e r i o u s limitations, obtained excellent results with their own form of ruby cylinder, giving the escapement new respectability in Europe. Consequently, modern cylinders evolved on the Continent, similar to the old English design but with better proportions and simplified methods for making escape wheels. These were used in some good watches but were most common in mediocre Swiss timepieces made in large quantities to sell at low cost, with resultant deterioration of the escapement's reputation. Use of cylinder escapements in America seems limited to a mere handful of pieces. At least one of the 17 Howard, Davis & Dennison 8-day watches was made with a cylinder, although the escapement may not have been made in the factory. Numbers 16 and 17 are marked "D. B. Fitts, Holliston, Mass." instead of Howard, Davis & Dennison, Daniel Bucklin Fitts being one of the first workmen at the Roxbury factory. In that period, Dennison and Howard could not always meet the payroll, so that Fitts may have taken compensation in watch material. His number 17 has a cylinder, though number 16 apparently does not, which Fitts could have put into the watch after he was no longer with the factory.[8] About the only other example was a model watch for the Cheshire Watch Co. of Cheshire, Connecticut. This was invented by Mr. A. E. Hotchkiss, who held numerous patents in the clock industry and owned land on which the Cheshire Watch Co. was built. In addition to a cylinder escapement, his model used lantern pinions with pins supported at one end only, and wheels soldered onto their hubs.[9] Since his design was rejected as impractical for production, Mr. Hotchkiss was not connected any further with the company, which then developed an inexpensive lever watch.

The duplex escapement evolved in France during the second quarter of the 18th Century, was patented in England by Thomas Tyrer in 1782, and was most commonly used in England. English makers favored it for high quality watches just below chronometers in status, till the lever escapement replaced it around 1850. As a high-grade jeweled watch the duplex required very careful construction, complicated by the fact that the escape wheel was difficult to make in the English form. Early French duplexes had two escape wheels on one arbor, like a Fasoldt escapement; one for locking and one for impulse, thus the name duplex. English makers combined these into one wheel with two sets of teeth, locking teeth extending radially from the rim and impulse teeth standing vertical. The teeth were pointed and delicate, so that cutting two sets on one wheel was a demanding job that required precision for a quality timekeeper. All this presented the duplex as an unlikely candidate for an inexpensive watch, yet two factors allowed the American industry to make vast use of it. First it was a frictional rest escapement, having only an escape wheel and balance, requiring no lever, detent, pallets or other parts. Secondly, much of the care and precision could be avoided if only mediocre performance was desired and it did not have to last many years. In devising his simple duplex for Waterbury longwind watches, D. A. Buck invented a form of escape wheel that could be punched in a press, and he set the balance wheel between flexible arms, cut out of the plates, that could be bent to adjust depthing of the escapement. The punched escape wheel merely had pointed radial teeth, with every other one bent upward to form an elevated impulse tooth. Even though the parts might wear out in a few years, replacements were cheaply available to repair the watch, which was sufficient for many millions of inexpensive timepieces.

One of the other makers of duplex watches in America was Buck's competitor, Jason Hopkins, who had designed

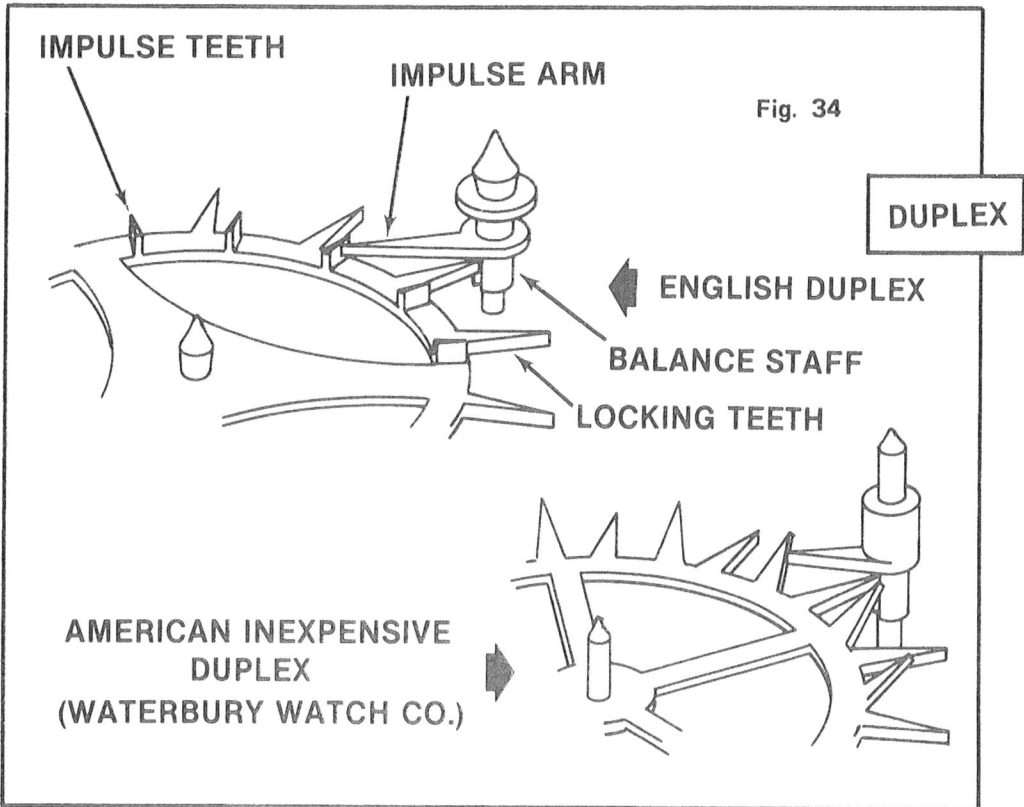

IMPULSE TEETH

IMPULSE ARM

Fig. 34

DUPLEX

ENGLISH DUPLEX

BALANCE STAFF

LOCKING TEETH

AMERICAN INEXPENSIVE
DUPLEX
(WATERBURY WATCH CO.)

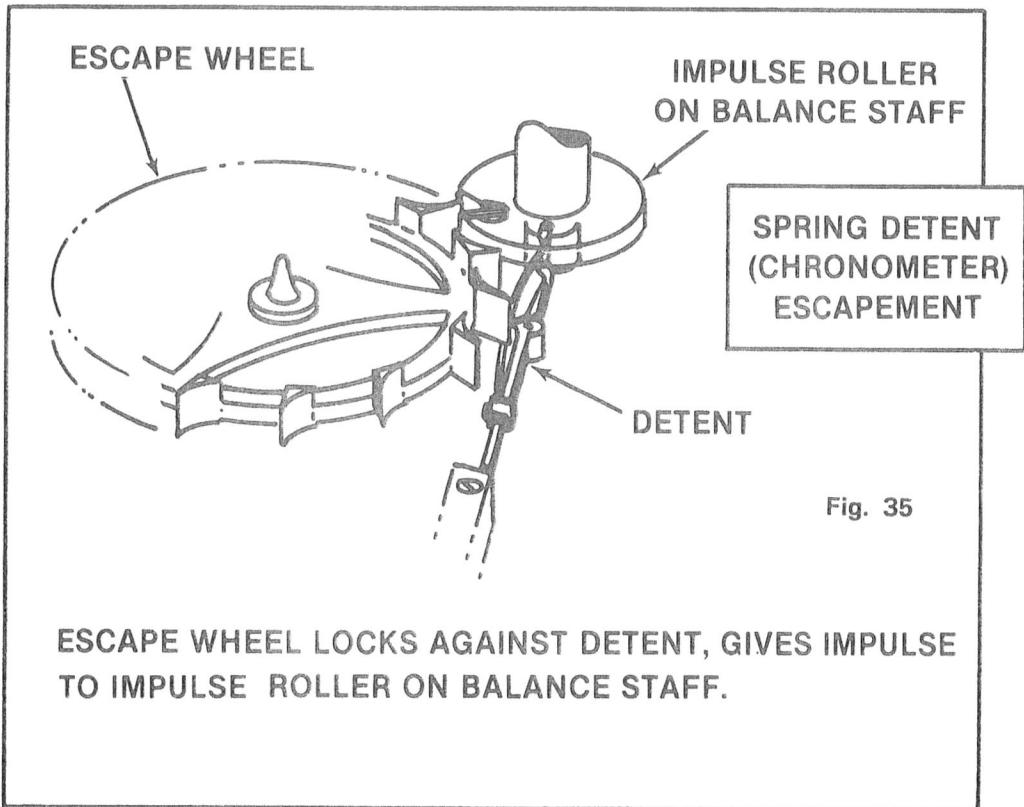

ESCAPE WHEEL

IMPULSE ROLLER
ON BALANCE STAFF

SPRING DETENT
(CHRONOMETER)
ESCAPEMENT

DETENT

Fig. 35

ESCAPE WHEEL LOCKS AGAINST DETENT, GIVES IMPULSE
TO IMPULSE ROLLER ON BALANCE STAFF.

179. Albert H. Potter, Geneva #588 ca. 1885 — lever escapement. In abstract fashion, the top plates form the initials A.P. Potter was perhaps the finest American watchmaking craftsman.

the Auburndale rotary as a cheap watch. Slightly before the Auburndale Watch Co. venture in 1876, Hopkins had formed the Washington Watch Co. in Washington, DC. In the 1874-75 period this group assembled some machinery and started about 50 movements, but the project was unsuccessful and no watches were completed. These were ¾-plate keywinders with a duplex escapement.[10] Little is known of the company or its product so that it is difficult to say now if these were high-grade watches of the English style or an inexpensive design as Hopkins tried to make at Auburndale. Albert H. Potter was the only other individual known to have made any quantity of duplex watches, which he built while living in the United States and Cuba. In 1876 he went to Geneva, Switzerland, where he concentrated more on levers and chronometers. All of Potter's work was of the highest order and the duplex watches would undoubtedly have been beautifully made.[11] The final source of American duplexes was the Suffolk Watch Co. of Waltham. In about 1901 they produced a small quantity of inexpensive movements using a duplex with machined escape wheel teeth, similar to English practice. Shortly thereafter, they were absorbed by United States Watch Co. of Waltham, and duplex production ceased.

The other escapement to be made in any quantity in America was the detent. Pivoted detent chronometers were invented by Pierre Le Roy in France about 1765, but the escapement developed more in England during the rest of the century through the efforts of John Arnold and Thomas Earnshaw. Instead of pivoted detents, in which the detent pivoted on an arbor set in jeweled bearings, the English used spring detents, with the detent firmly screwed to the plate but having a thin springy section allowing it to move. Swiss makers began using pivoted detents for high-grade watches in the 19th Century, but both in England and Switzerland use of chronometers was limited. Detent escapements were highly detached, had low unlocking resistance, and efficiently delivered impulse. Consequently they were used chiefly in high precision watches, from which connection they became known as chronometer escapements. While detents performed beautifully when well made, they were fussy to adjust and hardly worked at all

when carelessly constructed. The Swiss did make some detents of mediocre quality but these were scarce and virtually all chronometer watches were high grade.

American chronometer making was practiced by a fairly small group of individuals but they certainly produced more watches by far than the individually made cylinders and duplexes in this country. Interestingly, two very similar attempts were made by D. A. Buck and Jason Hopkins to use detent escapements in their inexpensive rotary watches. The idea did not get past the preliminary models in either case, and it's not clear what attracted these two men to detents. They were probably enticed by the simple appearance but were rapidly convinced that detent escapements were impractical for production. Though many chronometer makers likewise made but few examples, there were several who built a noteworthy quantity. One was George P. Reed, who had worked for both the Boston Watch Company and E. Howard during their early years and continued to work as a watchmaker in the Boston area for the rest of his life. During the 1860's he developed his pivoted detent, which used one piece as both the passing spring and recovering spring to return the detent to its resting position. Over his career he made perhaps 200 of these, also using his patented mainspring barrel that was licensed to Howard.[12] Another individual who reportedly made 200 chronometers was D. D. Palmer of Waltham, Massachusetts. For a number of years he was employed at the American Watch Co. in Waltham, also making watches at home on his spare time. Between 1860 and 1870 he made the pocket chronometers, which were 18 size, ¾-plate gilt watches, mostly with going barrels but a few with fusees. After 1870 he turned to making lever watches, eventually leaving his position at Waltham to practice his home watchmaking fulltime.[13] The other prolific maker was Albert H. Potter, who was mentioned previously in connection with duplex escapements. A number of the watches Potter made in New York used detent escapements, and after relocating in Geneva, chronometers were his chief product beside high-grade levers. Potter watches totalled around 600 by the end of his career, though it is difficult to say just how many were chronometers.[14]

Beyond these three individuals the quantity of American

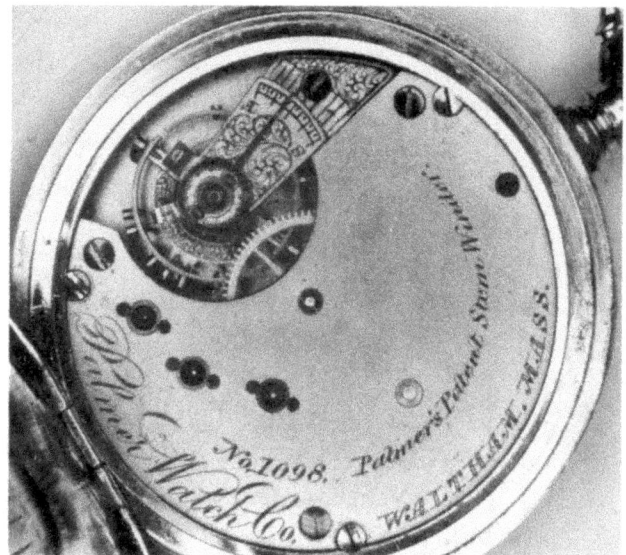

180. D. D. Palmer, Waltham #1098 ca. 1885 — Waltham model 68 material appears to have been used, finished by Palmer and combined with his stem-wind mechanism.

181. G. P. Reed, Boston #253 ca. 1875 — lever escapement with Reed's patented whiplash regulator and mainspring barrel that were licensed to Howard.

182. Dial of G. P. Reed #253 — showing up-down indicator and beautiful matching hands.

chronometers drops off rather rapidly. There was a J. H. Allison who made about 30 watches in the Detroit and Chicago areas beginning in 1853. Back east in Newburyport, Massachusetts, Norman Greenough made around 15 pocket chronometers before his death in 1866, and an Albert E. Potter made a few unusual watches in Boston with complicated escapements on the detent principle. Otherwise there were individuals who made only a few chronometers, such as Jacob Karr of Washington, DC,[15] and Jonas Hall of Roxbury, Vermont.[16]

The final area of escapement variation was the hairspring, where a number of designs appeared and disappeared during early years of the industry in attempts to make isochronous springs. Isochronous operation meant that the balance had the same frequency whether it swung in large arcs or small arcs, thus being insensitive to changes in motive force. This was an important consideration in eliminating the fusee, which kept motive force constant as the mainspring ran down. The most common hairspring variations were helical, or cylindrical, springs since they showed some promise of being manufacturable in quantity.

English chronometer maker John Arnold had invented helical hairsprings, with properly adjusted terminal curves to achieve isochronism and these were universally used in English chronometers thereafter. E. Howard & Co. tried using helical springs in their first year of operation, about 1858, since precision watches were Howard's objective. Though it was rumored that only three of these were made, some slightly greater quantity was produced. Cylindrical springs resulted in thick watches, and probably did not turn out to be that easy to manufacture anyway, so the experiment was soon discontinued. Interestingly, one of the Howard helicals was marked "ISOCHRONON" on the dial, clearly indicating what was intended.[17] A few years later the American Watch Co. in Waltham advertised watches with helical springs and special adjustments, but it is not known if any were made. In some of his early watches Charles Fasoldt used an odd variation of the helical spring with a flat spiral wound across the bottom.

These were used in some English watches and were usually referred to as a duo-in-uno spring. About the only other use of helical balance springs in American watchmaking was by chronometer makers, who were not hampered by quantity production considerations, and used them as the traditional and proven method of obtaining quality performance.

A unique attempt at making an isochronous balance spring was the vibrating hairspring stud used by the American Watch Company. After purchasing the Nashua Watch Co. in 1862, American marketed ¾-plate keywind watches, based on the Nashua design, as their highest grade model. These could be ordered with two interesting options. One was the mainspring barrel patented by Nelson Stratton to protect the watch from damage in the event of a broken mainspring, and the other was Fogg's vibrating hairspring stud. Charles W. Fogg, superintendent of the ¾-plate department, patented the safety pinion used by Waltham as well as his vibrating stud. With this latter design, the outer end of the hairspring was fastened to a pivoted arm which rocked back and forth as the hairspring expanded and contracted. The idea was invented by chronometer maker William Hardy of Clerkenwell, England, in 1806.[18] Instead of a pivoted arm, Hardy fastened the top coil of a helical spring to a short length of slender straight spring planted to the balance cock, which achieved a similar result. Such elastic studs appeared with flat hairsprings on both marine and pocket chronometers by Parkinson & Frodsham of England, and as late as 1900 on a chronometer watch by Bridgeman and Brindle. This latter example used two such hairsprings, one above the other, wound in opposite directions.[19] It can be seen that the idea was taken seriously by English chronometer makers for a number of years, and English watches were likely the inspiration for Fogg's design. Nelson Stratton had been sent there around 1852 by Dennison and Howard to learn the art of gilding, and being an observant mechanic he would have learned much more before returning home. Also, English chronometers were available in the United States, since England was the foremost supplier of chronometers, so that the elastic stud was known to some in the general trade. The American

Watch Co. used vibrating hairsprings to achieve isochronism by simple means. A pivoted arm was easier for them to make than the elastic spring design used in England. These did not really work, however, and were discontinued shortly.

In the end, early American watch companies abandoned schemes such as helical hairsprings and vibrating studs, and used imported Breguet springs for their high grade watches. Between 1860 and 1880 the industry developed techniques for making Breguet springs hardened and tempered in their final form. It seems interesting that old ideas die slowly, for in the 1930's Hamilton Watch Co. made one final attempt at a resilient hairspring stud. On a 992-E movement they used a hairspring with the last coil extended into a straight section along the side of the balance cock, which had a screw driven curb for regulating mean time. This not only gave the effect of the old resilient stud, but was almost identical to the ancient Barrow regulator (named after Nathaniel Barrow) used on the earliest balance spring watches around 1680.

Just as Breguet overcoils replaced other designs for achieving isochronism, the conventional lever escapement replaced all other types in the American factory system, with the obvious exception of inexpensive duplex watches. No other design could be constructed so easily by factory methods and obtain quality performance in hard daily use. These were clearly the same factors which made levers universal, even in the handcraft systems of England and Switzerland. After centuries of evolution the detached lever escapement emerged as the most simple yet adequate design, by unanimous agreement, and the early years of the American watch industry saw the last unsuccessful challenges to that title swept away.

REFERENCES

1. Charles S. Crossman, *The Complete History of Watchmaking In America* (reprinted Exeter, NH, Adams Brown Co.), p. 5.

2. Frederick Mudge Selchow, *Belding Dart Bingham — The Nashua Watch Company* (BULLETIN, National Association of Watch and Clock Collectors, Inc., December 1975), p. 539.

3. W. J. Gazeley, *Clock and Watch Escapements* (Newnes-Butterworth, London, England, 1975), p. 188.

4. *Pocket Timepieces of the New York Chapter Members* (New York Chapter, NAWCC, 1968).

5. Paul M. Chamberlain, *It's About Time* (New York, Richard R. Smith, 1941), p. 101.

6. Ibid., p. 104.

7. Ibid., p. 69.

8. George V. White, *Mr. Edward Howard's Watch No. 1* (BULLETIN, National Association of Watch and Clock Collectors, January 1945), p. 35.

9. Crossman, p. 165.

10. Ibid., p. 169.

11. Ibid., p. 205.

12. Ibid., p. 191.

13. Ibid., p. 187.

14. Chamberlain, p. 444.

15. Frederick Mudge Selchow, *Jacob Karr — Unlisted Watchmaker* (BULLETIN, National Association of Watch and Clock Collectors, Inc., December 1970), p. 812.

16. A. Dodge & G. Lucchina, *Jonas G. Hall, 1822-1891* (BULLETIN, National Association of Watch and Clock Collectors, Inc., October 1976), p. 436.

17. Thomas L. De Fazio, *The Nashua Venture and the American Watch Company* (BULLETIN, National Association of Watch and Clock Collectors, Inc., December 1975), p. 586.

18. Rees, *Clocks, Watches and Chronometers* (reprinted by Charles E. Tuttle Co., Rutland, Vermont, 1970), p. 153.

19. Gazeley, p. 250.

X
INDIVIDUAL MAKERS

American watchmaking began before the Revolution as an individual handcraft, and individual watchmaking was practiced to some degree throughout the years of the industry. Very little information is available on those early makers who worked before Luther Goddard opened his shop in 1809. Their watches would look like English pieces of the period, and indeed many of the parts would have come from England. Adding to the confusion were importers and jewelers who engraved their names on foreign watches, such as Effingham Embree of New York. Identifiable American watchmakers appeared as the 19th Century progressed, producing pieces which were unmistakably theirs, even though some of the parts such as balances, springs, jewels and wheels may have been imported or bought from the industry. Luther Goddard was one of the earliest and most prolific of such makers, turning out about 530 watches between 1809 and 1817. He certainly was one of the most important single makers and employed several apprentices who went on to make watches of their own. One was Jubal Howe, who probably influenced Aaron Dennison before the days of the industry, and another was James Hamilton, who reportedly made 80 watches over his career. These people worked mostly before Dennison's Boston Watch Co. brought about industrialized production, and only one other well documented individual made watches before the industry began. That was Jacob D. Custer (1805-1872), a sort of mechanical genius around Norristown, Pennsylvania. He had grown up with little formal education near the farm of David Rittenhouse, being self-taught and becoming well informed as he pursued various mechanical interests throughout his life. About 1831 he began producing clocks, making tower clocks in iron and steel as well as hall and shelf clocks with metal movements. Between 1841 and 1842 he also produced a dozen watches, about 14 size and ¾ plate with fusees and jeweled lever escapements, making the movements and cases himself except for hairsprings and fusee chains.[1] Later he made lighthouse clockworks for the government and bullets during the Civil War using a machine of his own design.

Custer's watchmaking took place before factory-made watches were available, but even as mechanized watchmaking began to grow there were a few who sought a living as individuals in the field. One of the most notable of those was Charles Fasoldt (1818-1898) who came to America from Germany in 1848 because of civil strife in that country. He had planned to move to one of the western states here but stopped in Rome, New York, to see his brother and was soon preparing to manufacture tower clocks there. By 1850 he was making watches, apparently following the occupation of his father, and eventually patented a regulator, hairspring stud and the escapement he used in his watches and clocks. The patent regulator, which utilized an adjustment screw, was the subject of a successful lawsuit against Edward Howard when Howard began using G. P. Reed's whiplash regulator. In 1861 Fasoldt moved his small operation to Albany and through 1880 produced about 550 watches. While he hired helpers to make his watches, it was also a family affair. His son reported that Fasoldt made the wheels, arbors, plates and hairsprings while his daughters made jewels, patent hairsprings and patent regulators. Though many of his watches were similar they were all distinctive in some manner, exhibiting indivduality and interesting variations. There were helical hairsprings, non-rusting hairsprings of gold alloy, some pocket chronometers, double-train timers, ¾-plate movements and a somewhat Swiss-style bridge layout on his later watches.

Most of his pieces were keywind, but approximately the last 150 were stemwind with the curious feature that setting the hands still required a key, inserted from the back, as with some Empire City watches made in Marion, New Jersey. At the Philadelphia Centennial Exposition Fasoldt entered a small tower clock, which took the highest prize for tower clocks exhibited, and in his later years took up microscope making and ruling of diffraction gratings.[2]

During this same period, 1850's through the 1880's, J. H. Allison made about 30 watches in Detroit and Chicago. Most of these were chronometer escapements, both key and stemwind, some with fusees. They were made in fits and starts over the years while Mr. Allison worked at jobs in the jewelry trade, eventually going into the retail jewelry business himself in Indiana.[3]

Also during the 1850's began the career of the finest of all American watchmakers, Albert H. Potter (1836-1908). He had been born in Saratoga County, New York, and apprenticed in Albany, tempting speculation that he may have crossed the path of Charles Fasoldt, though it is not likely. In 1855 he established himself in New York City, staying there for six years during which he made about 35 movements that sold in gold cases. These were keywind lever and chronometer escapements, mostly going barrel but some fusee. He then went to Cuba for several years where he continued watchmaking, including a quarter repeater and duplex escapement, but returned to the United States to form a company in Chicago with his brother in 1872. This apparently did not satisfy him either for in 1876 he relocated in Geneva, Switzerland, where he made about 600 watches, developing a number of escapement variations and special tools for accurately making escapements. It was in Geneva that Potter established a reputation of the highest order for which he was known long after his death. Because of his reputation Potter had no trouble obtaining work from the finest craftsmen of Geneva, which he combined into beautifully designed and executed timepieces that were exquisite masterpieces of the watchmaker's art. These included many levers and chronometers as well as complicated watches, with prices that began at $250. Also he invented the "Charmilles" watch made in that section of Geneva around 1894. This was an attempt to make a quality inexpensive watch, based on clever but serious design, but few were ever produced.[4] While it was unfortunate that Potter felt he could not stay in America to practice watchmaking his way, he may have been right. At the time, American factories had established good reputations with the general public, especially in the line of popular grade watches. For watches of the utmost quality, wealthy individuals still looked to the finest makers of England, plus high grade Swiss watches by such people as Jurgensen, Ekegren, Nardin and Patek Phillipe. It just may have been that in Switzerland Potter found a better climate in which to build unusually fine handmade watches, as well as finding that the Swiss name helped in selling watches of that class.

The Boston area had a number of lone watchmakers. Coincidentally, one was named Albert E. Potter, who worked for both Waltham and E. Howard. While in their employ he made several watches with complicated escapements based on the chronometer. These were reputedly quite skillfully made. Later he abandoned this recreation and went into the retail jewelry trade in Boston.[5] During the early years of the industry the city of Boston also had a jeweler named Hiram Smith who attempted to make watches around 1860, both bridge and fullplate models. It was stated that he constructed his watches with interchangeable parts. Such an effort would indicate that he was serious about watchmaking, but competition from imports and factories, especially right around Boston, made it difficult to stay in business on a small scale.[6] A small-scale watch manufacturer who did manage to support himself for a number of years was D. D. Palmer of Waltham. He had come to the Waltham factory from New York state where he had run a retail jewelry store, and while in New York had made about 25 chronometer watches. At Waltham he continued to make pocket chronometers till perhaps 200 had been completed, and finding that he preferred making his own watches he left the factory in 1875 to set up a shop in his nearby home. From his home he also operated the first watchmaking school in the country. It was reported at the time that he was working on material for 1500 watches, all lever escapement, 16 size and ¾ plate in both gilt and nickel, and employing his own stemwind design.[7] Some of these were completed but it is not known just how many or for how long Mr. Palmer remained in the business.

Another individual watchmaker in the Boston area was George P. Reed, who patented the stationary barrel and whiplash regulator used in E. Howard watches. Reed had come to the old Boston Watch Co. in 1854, continuing till the change in management in 1857 when he went with Howard back to the Roxbury factory and remained till 1865. Shortly after leaving Howard he designed his own chronometer escapement while supporting himself with a retail shop in Boston. Next he moved to the Boston suburb of Malden where he built about a hundred watches over a three-year period. He then occupied the old quarters of the Tremont Watch Co. in nearby Melrose and remained in that town while continuing to make watches. His timepieces were nearly all a sort of ½ plate and 18 size, using both levers and his chronometer as well as his patent barrel. These numbered about 500 in all, 200 chronometers and 300 levers, and he managed to support himself as a watchmaker for the remainder of his life by finding retail outlet for his watches in Boston.[8]

One of the most romantic individuals in American watchmaking was Don Mozart (1820-1877) whose family immigrated to Boston from Italy. As a child he was kidnapped aboard a departing vessel while playing on the Boston waterfront and never saw his family again. He found his way back to the United States, wandering about for a number of years working at mechanical occupations and finally taking up watchmaking, which had been his father's occupation. He lived in Ohio after marrying during the 1850's, but returned east in 1863 to attempt manufacturing a clock of his design that ran for one year. 1863 was the one year it ran, for the project soon failed, but in 1864 Mozart interested a group of New York backers to form the New York Watch Company. They established him with a shop in Providence, Rhode Island, to begin work on his "three wheeled

183. Cut of Mozart watch made in Ann Arbor, Michigan, ca. 1870 — balance rim has weights rather than screws, and Mozart's unique escapement and three-wheeled train are used.

184. C. H. Hoyt, Detroit #54 ca. 1872 — ¾ plate lever watch made for M. S. Smith jewelers. Smith movements sometimes utilize material of mixed serial numbers.

watch" but after two years there was only some machinery and no watches. The backers removed Mozart in favor of a more conventional workman from Waltham to build a ¾-plate American watch, and the operation eventually became the Hampden Watch Co. of Springfield, Massachusetts. Mozart immediately went to Ann Arbor, Michigan, where in 1868 he formed the Mozart Watch Co. to make his watch. This company also failed, but not before a small quantity of watches was presented to financial backers in the scheme.[9] Mozart became mentally feeble in his later years and died near Ann Arbor after a complete breakdown. While he operated through formation of companies, he was their prime mover and their products always reflected Mozart's sense of invention. In fact, he may have led a more successful career had he produced and sold his watches as an individual maker, for the escapement didn't lend itself to factory production, nor would the jewelry trade have found it a welcome piece on which to attempt repairs. Interestingly, the Rock Island Watch Co. was formed just after failure of his Ann Arbor venture, independently of Don Mozart, to make a watch on the Mozart design, but also collapsed before any watches were made.[10]

Just after 1870, Mr. C. H. Hoyt was making lever watches in Detroit bearing the name of a prosperous jewelry firm, M. S. Smith & Co.[11] These were a quality nickel-finish ¾-plate movement in a standard 18 size, typical of the style becoming fashionable in American high-grade watches. They were transition movements having stemwinding with lever set, but also having winding and setting keys in the standard locations. Machinery was used to make interchangeable parts, and many Hoyt movements were assembled from mixed serial number plates. This operation sufficiently attracted a Detroit industrialist named Eber B. Ward that Ward purchased Mr. Hoyt's shop, as well as the old Mozart machinery from the Rock Island Watch Company, which had never used it. Hoyt and the machinery were installed during 1875 in a new building in Freeport, Illinois, as the Freeport Watch Co. to make watches of Hoyt's design. A couple dozen watches were completed under the Freeport name, using Hoyt's old stock, but the building soon burned with all contents, ending the Freeport

Watch Company. Between Smith and Freeport, about 100 Hoyt watches were completed.

Washington, DC, was the home of several individual watchmakers. One was Jason Hopkins, designer of the Auburndale rotary watch, though he never became actively involved with that company. He had previously made a few fine quality movements by hand in the Washington, DC, area. Desiring larger enterprise, he started the Washington Watch Co. there, managing to complete some machinery and about 50 movements with duplex escapements. This project absorbed its available cash and simply ceased, after which Mr. Hopkins became taken with his rotary design.[12] Another Washington watchmaker was Jacob Karr. In 1864 and 1865 he took out patents on his unusual form of pivoted detent chronometer, and he was listed on several chronograph patents in 1882.[13] Few examples of his watchmaking survive and little is known of his life.

185. Freeport Watch Co. #11 ca. 1875 — Hoyt material finished in Freeport, Illinois.

Much more is known about Ezra Bowman of Lancaster, Pennsylvania. In 1879 he opened a retail jewelry store in that town where he also attempted to make a high quality watch. For this purpose he hired Mr. William Todd, an Englishman who had previously been with E. Howard, Elgin and the Lancaster Watch Company. The movement was a ¾-plate, free-sprung design modeled after Frodsham watches of London and using an English side lever escapement with pointed escape teeth. They purchased dials and imported fine English balances but all other parts were manufactured in their own shop on some of the best machinery from the defunct United States Watch Company. Material for about 50 watches was begun and some of these were completed, but after two years of effort the operation was detracting excessively from the retail store and Mr. Bowman consequently sold the equipment.[14] The buyer was Josiah P. Stevens of Atlanta, Georgia. Stevens was a bright and ambitious man from a learned family but due to the chaos of the Civil War he had received little formal education. He had gone to Atlanta, becoming proprietor of that city's largest jewelry establishment, and in 1882 obtained the Bowman machinery to open a separate watchmaking business. He also bought additional equipment from the

186. Ezra F. Bowman, Lancaster, PA, #19 ca. 1880 — lever watch with free-sprung balance and English-style pointed-tooth escape wheel.

Stark Tool Co. of Waltham, such as screw machines, staff lathes, wheel and pinion cutters, etc., and hired the services of a Mr. Bagley of Waltham, who retured with Stevens to make dies for punched parts like wheels, levers and setting mechanisms. William Todd, who had done much of the work on the Bowman watches, also went to Atlanta to help design Stevens' watch, which emerged as a 16-size, ¾-plate movement with club-tooth lever escapement and Stevens' patented regulator. This consisted of a rotating gold disk in which a spiral slot was machined that carried the end

187. J. P. Stevens, Atlanta, GA, #553 ca. 1885 — Stevens made a ¾-plate lever watch using Ezra Bowman's tools. However, this particular movement is made with material from New York Watch Company's E. W. Bond model. Note Stevens' spiral regulator.

88 86 84 82 80 78 76 74 72 70 68 66 64 62 60 58 56 54 52 1850 48

INTO THE 1890'S

JONAS G. HALL

CHARLES FASOLDT

J. H. ALLISON

ALBERT H. POTTER

HIRAM SMITH

LYMAN THOMPSON

NORMAN GREENOUGH

DON MOZART

JACOB KARR

D. D. PALMER

G. P. REED

JASON HOPKINS

WASHINGTON W. CO.

HOYT

FREEPORT W. CO.

JOSEPH DELPHIN

ALBERT E. POTTER

J. P. STEVENS
(ATLANTA, GA.)

EZRA BOWMAN
(LANCASTER, PA.)

HERMAN VON DER HEYDT

Fig. 36

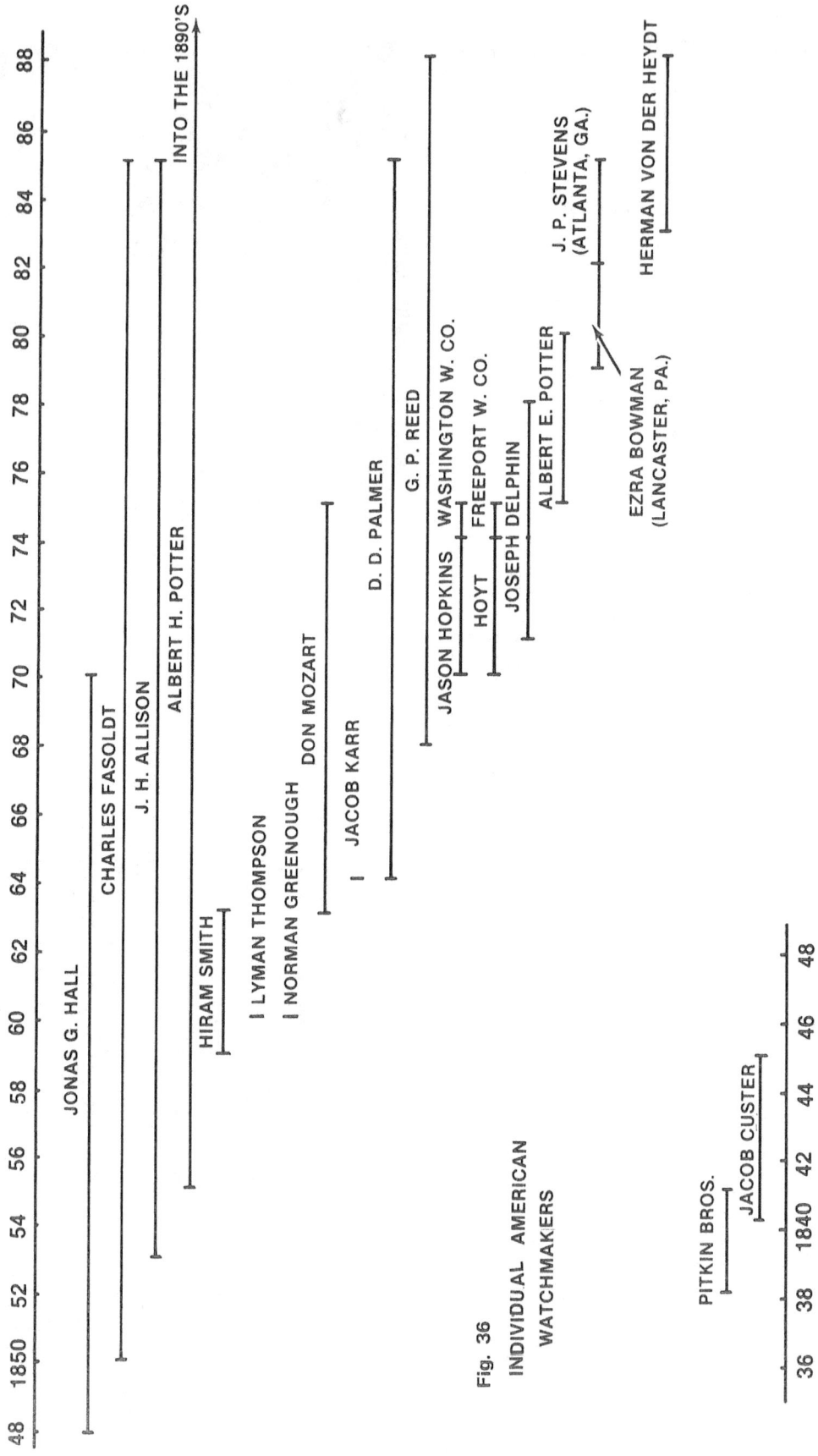

INDIVIDUAL AMERICAN
WATCHMAKERS

PITKIN BROS.

JACOB CUSTER

36 38 1840 42 44 46 48

1981 © M. C. HARROLD

122

188. Nowman Saati, Providence, Rhode Island, ca. 1895 — one-wheeled watch design in marine chronometer size. The escape wheel is in the center. Arms extending to the center from next to the mainspring barrel power a worm gear on the center shaft.

of a standard regulating arm. Fewer than 200 movements had been made by 1885 and since a number of workmen were Yankees who desired to return north during the hot summers, Stevens decided to close the shop. The company continued to remodel Hampden and Longines movements with the patent regulator and mark them with the company name, but no other original watches were made.[15]

One of the remaining individuals who made any quantity of watches was Herman Von Der Heydt, who came to this country in 1881 from Wiesbaden, Germany, where he had been trained as a watchmaker. By 1884 he had patented his selfwinding watches and began producing them intermittently along with running a jewelry and repair business in Chicago. These were 18-size, lever watches, fully jeweled and of nice quality. Selfwinding was accomplished by a pendulum type of winding weight which could swing through a limited arc and covered about ¼ of the pillar plate. Von Der Heydt received offers from watch companies and jewelers to take up manufacture of the watch on a larger scale, but preferred to make them at his own pace, finishing perhaps 40 of them over the years with only a minimum of tools.[16]

Late in the century came an individual with one of the most unusual ideas in watch construction. Nowman M. Saati of Providence, Rhode Island, received a Swiss patent in 1894 for what he called his "wonderful one wheel watch." He listed himself as a former watch manufacturer of Chaux-De-Fonds, Switzerland, and reportedly had plans for establishing a shop in Providence for making his watch. It is unlikely that more than a few samples were ever completed. The only conventional gear in this mechanism was a large escape wheel that rotated once per minute. As with New York Standard's worm drive, this escape wheel had a worm machined into its arbor rather than a standard pinion. The worm was driven alternately by either one of two arms pivoted between the plates. As one arm drove the length of the worm, the other arm dropped from the exit end of the worm back to its beginning. These arms were

actuated by heavy teeth in the mainspring barrel, which functioned as cams rather than gearteeth. Mister Saati's claimed advantage for this invention was simplicity. In reality, the cam action and sequencing to operate the arms were complicated and unreliable, especially in watch size, and escape wheel torque was variable. Some samples were made to operate, and the design received an award at the Geneva Horological Exhibit of 1896. On the other hand, the design did not revolutionize watch manufacture.

There were other individuals who made watches in small quantities such as Jonas Hall of Roxbury, Vermont, E. H. Flint of Cincinnati, and E. C. Dwight of Freeport Illinois. Some of their stories can be found in available literature and occasional watches can be located. Because such people were often obscure, and their watches can be mistaken for imports, many will remain unidentified. Their lives and work are certainly not without importance and much more research is required in this area since these makers were nearly always more interesting than the mechanical and repetitive operations of the factories. Even in the land where watchmaking became an industry, there always remained individuals who maintained watchcraft as a personal enterprise.

Nevertheless, the mainstream of American watchmaking was the industrial manufacture of interchangeable parts, though it should not be interpreted merely as a story of brick buildings, steam engines and automatic lathes. It was all brought about by people who were equally as unique and interesting as the individual handcrafters. The Pitkins, Aaron Dennison, and Edward Howard were inventive watchmakers who placed themselves firmly in horological history. There was no industry when they began and what they accomplished sprang from their own ideas and initiative, ranking them with the most adventuresome personalities in watchmaking. Others that followed, like Nelson Stratton, J. C. Adams and James Gerry, were creative individuals who wove notable personal histories while shaping the character of the American watch industry. As with every human endeavor, watchmaking, even on an industrial scale, is a story of people.

REFERENCES

1. S. H. Barrington, *Custer and His Clocks* (BULLETIN, National Association of Watch and Clock Collectors, October 1949), p. 683.

2. Paul M. Chamberlain, *It's About Time* (New York, Richard R. Smith, 1941), p. 428.

3. Charles S. Crossman, *The Complete History of Watchmaking in America* (reprinted Adams Brown Co., Exeter, NH), p. 200.

4. Chamberlain, p. 444.

5. Crossman, p. 204.

6. Ibid., p. 201.

7. Ibid., p. 187.

8. Ibid., p. 191.

9. Chamberlain, p. 225.

10. Crossman, p. 119.

11. Ibid., p. 121.

12. Ibid., p.169.

13. Frederick Mudge Selchow, *Jacob Karr — Unlisted Watchmaker* (BULLETIN, National Association of Watch and Clock Collectors, Inc., December 1970), p. 812.

14. Crossman, p. 170.

15. W. B. Stephens, *J. P. Stevens — His Watch Company and Inventions* (BULLETIN, National Association of Watch and Clock Collectors, December 1954), p. 321.

16. Henry G. Abbott, *Watch Factories of America* (reprinted Adams Brown Co., Exeter, NH), p. 137.

APPENDIX A
AMERICAN WATCH COMPANIES

Appendix A lists American watch manufacturing companies existing from 1840 to 1930. If a company fails to appear, it is likely for one of the following three reasons:

1. Individuals are not listed. Only companies and organizations are given in the Appendix. Individuals are found in Chapter X, Individual Makers.

2. Only true manufacturers are listed. Contractors and model names are not included. Many organizations contracted for watches to be made in their name, such as retail jewelers, wholesale jewelers, sales companies and catalogue businesses, which are not included here. Also some names, such as Home Watch Company, are model names used by watch factories. Several notable contractor and model names have been listed for convenience. Otherwise an identification guide book will help decipher such watches.

3. Only American manufacturers are listed. An unlisted company may simply be a legitimate foreign producer. Also many contractors used foreign sources. Thus an American name may appear on a Swiss watch, and some Swiss watches were very American in appearance. Even high-grade Swiss watches used American split plate movement styles and standard sizes, especially after the railroad standards were issued. Finally, there were Swiss fakes. These were usually inferior quality watches with American sounding names, designed to fool gullible customers. On many of these the name or location is misspelled, apparently intentionally.

Abbott Watch Co. — name used by the Keystone Howard Watch Co.

Adams & Perry — see Lancaster Watch Co.

American Horologue Co. — see American Watch Co.

American Waltham Watch Co. — see American Watch Co.

American Watch Co.[1,2]

1850	Howard Davis & Dennison	Boston, MA
1851 (6 mos.)	American Horologue Co.	Boston
1851-1853	Warren Mfg. Co.	Boston
1853-1857	Boston Watch Co.	Waltham, MA
1857 (1 mo.)	Tracy Baker & Co.	Waltham
1857-1859	Appleton Tracy & Co.	Waltham
1859-1885	American Watch Co.	Waltham
1885-1906	American Waltham Watch Co.	
		Waltham
1906-1923	Waltham Watch Co.	Waltham
1923-1925	Waltham Watch & Clock Co.	
		Waltham
1925-ca. 1957	Waltham Watch Co.	Waltham

The family of companies represented by the American Watch Co. ranged from the first efforts of Dennison and Howard to the giant "Waltham" plant. All grades of conventional jeweled watches were manufactured within the companies. Their lowest priced models were never the worst available and their finest were admirable quality. These companies were always pioneers, developing the American fullplate and ¾-plate watch, improvements such as stemwinding, attachments like chronographs and repeaters, and highly sophisticated machinery and factory methods. As a result, their ideas and people went forth to build much of the remaining industry and the variety of Waltham products is immense.

189. Swiss fake of Waltham P. S. Bartlett — note that the movement is marked P. S. Barton.

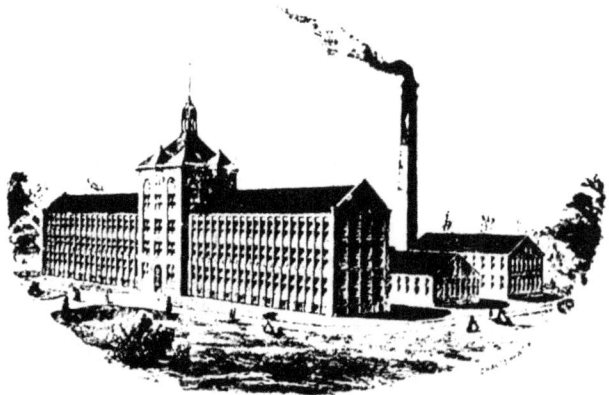

190. Factory of the Aurora Watch Co., Aurora, Illinois, ca. 1885.

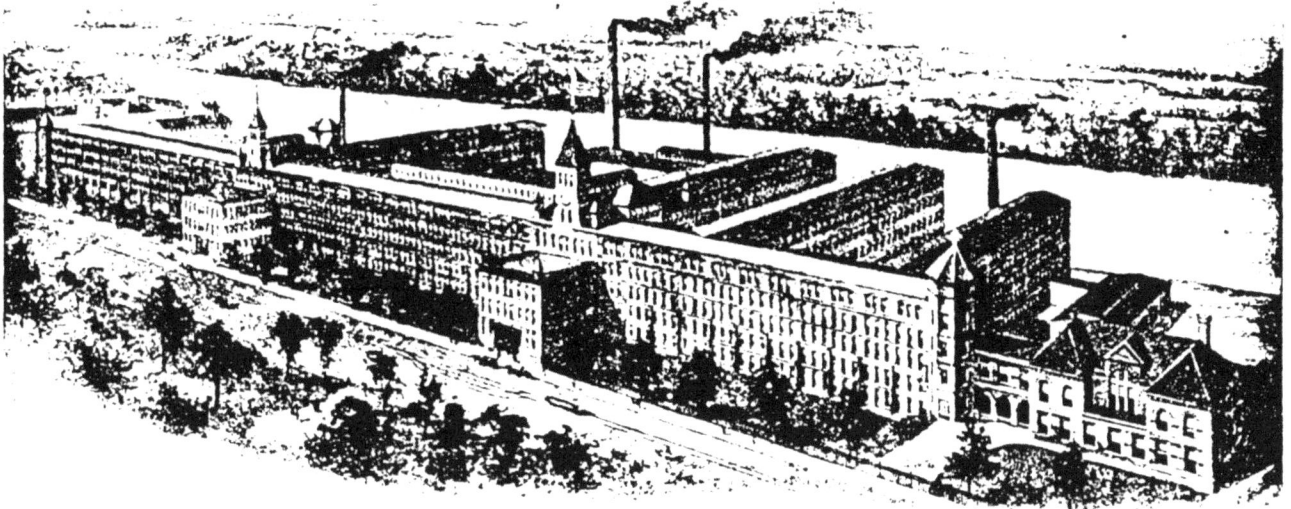

191. The Waltham Factory — American Waltham Watch Co. 1904.

Ansonia Clock Co. — 1850-1930, Ansonia, CT, began watch-
    making in 1904.

Ansonia was one of the moderately-sized producers of
dollar watches. The factory was bought by Russia, along
with Hampden Watch Company, to establish the watch industry in that country.

Appleton Tracy & Co. — see American Watch Co.

Appleton Watch Co. — 1901-1903, Appleton, WI

The Remington Watch Company, organized by O. E. Bell,
purchased machinery of the defunct Cheshire Watch Company. They made a watch similar to the old Cheshire, plus
cases, selling complete watches marked Appleton Watch
Company.

Auburndale Watch Co. — 1876-1883, Auburndale, MA[3]

This company built the Auburndale rotary, first attempt
at a "dollar watch" type of timepiece, which failed due to
faulty design and poor equipment. They also tried making
an inexpensive jeweled watch which failed. The company
continued a brief while making timers and metallic thermometers.

Aurora Watch Co. — 1883-1886,* Aurora, IL

Aurora was a brief attempt to make fullplate railroad
watches in the midwest before the railroad watch standards
were established. They produced good watches, some with
a "fifth pinion" for the second hand. Their machinery was
sold to the founders of Hamilton.

Ball Watch Co.[25] — 1891-present, Cleveland, OH

Webb C. Ball, principal personality behind the railroad
watch standards, set up a contracting company to sell railroad and non-railroad watches. Hamilton and Waltham
were his main suppliers but watches were also obtained
from Elgin, Howard, Hampden, Illinois, and Seth Thomas.
Ball patented plate designs and registered his trade marks;
his watches were all made to his own standards and designs.

Bannatyne Watch Co. — see Ingraham.

The company's demise may be as late as 1890.

192. Ball Watch made by Waltham ca. 1905 — nearly
identical to the Illinois Watch Co. version of this design
(illustration 130).

193. Columbus Watch Co. #23,734 ca. 1884.

Benedict & Burnham — see Waterbury Watch Co.

Boston Watch Co. — see American Watch Co.

Bowman, Ezra[19]

1879-1882	Ezra Bowman	Lancaster, PA
1882-1885	J. P. Stevens	Atlanta, GA

Ezra Bowman tried small-scale manufacture of a high-grade watch, using some of the best equipment from the defunct U.S. Watch Co. of Marion, New Jersey. This overburdened his retail jewelry business and he sold the equipment to J. P. Stevens of Atlanta. Stevens expanded the operation and tried to produce a less exotic watch but could not keep the business alive.[19] Stevens later remodeled Hampden and Longines movements with his patent regulator.

Burlington Watch Co. — name used by a sales company on watches made by Illinois.

California Watch Co. — see Cornell Watch Co.

Cheshire Watch Co. — 1883-1894, Cheshire, CT

This was one of the earliest of inexpensive jeweled watchmakers. They used a combination of stamped and machined parts to produce a jeweled movement at low cost. Machinery was sold to the Remington Watch Company of Appleton, Wisconsin, in 1901.

Columbia Watch Co.[20] — 1896-1901, Waltham, MA

Columbia was founded by Edward Locke, who began the Waterbury Watch Company. The firm made inexpensive jeweled watches, mostly of small size, and after becoming Suffolk Watch Co. in 1901 made duplex escapement watches. These used a machined escape wheel slightly different from the Waterbury design. They were absorbed by United States Watch Co. of Waltham in 1901, which was soon bought by Keystone Watch Case Company.

194. Columbus Watch Co. factory ca. 1885.

Columbus Watch Co.[5]

| 1882-1903 | Columbus Watch Co | Columbus, OH |
| 1903-1929 | South Bend Watch Co. | South Bend, IN |

This was organized out of the Gruen and Savage watch importing business of Columbus. They made railroad watches before the standards plus moderate quality watches. The entire plant was purchased and moved to Indiana to establish the South Bend Watch Co. which made good quality railroad watches.[18]

Cornell Watch Co.[13]

1864-1870	Newark Watch Co.	Newark, NJ
1870-1874	Cornell Watch Co.	Chicago, IL
1874-1876	Cornell Watch Co.	San Francisco, CA
1876 (6 mos.)	California Watch Co.	Berkeley, CA
1880	Western Watch Co.	Chicago, IL

Newark Watch Co. was one of the rush of new companies in 1864 making gilded fullplate movements similar to the Boston watch. When it failed John C. Adams and Paul Cornell organized the Cornell Watch Co. as a land development scheme in Chicago, utilizing machinery from the defunct Newark company. This soon failed and the San Francisco venture was hastily assembled as an escape, which quickly failed in turn. Some material was finished under the name California Watch Co. and the remaining material was finished by a former foreman, Albert Troller, in Chicago as the Western Watch Company. All along the watches being made looked like the old Newark. Machinery was finally sold to the Independent Watch Co. of Fredonia, New York, which was then reorganized as the Peoria Watch Co. and still made a watch of the Newark design.

Dennison Howard & Davis — name used by the Boston Watch Co.

Dudley Watch Co.[4]

1920-1925	Dudley Watch Co.	Lancaster, PA
1925-1929	P. W. Baker & Co.	Lancaster
1929-1935	J. W. Apple Co.	Lancaster
1935-1968	X-L Watch Co.	New York, NY

Mister William Wallace Dudley formed the company to manufacture pocket watches in which bridges took the form of Masonic symbols. Dudley made plates and barrels while wheels, balances, regulators, and bits of setting mechanism were bought from Waltham, Hamilton, and Swiss special orders. Sufficient enthusiasm was not to be found among Masons to support a watch company, compounded by rapidly growing popularity of wristwatches in the 1920's. Baker and Apple were a subsequent receiver and purchaser respectively of the company, neither of whom met with particular success in the Masonic watch business. X-L Watch Co. moved Dudley machinery and inventory to New York, finishing approximately 1000 movements from old stock over the next 35 years. Total production probably reached about 3500 during a half-century of sporadic manufacture.

195. South Bend Watch Co. factory ca. 1905 — filled with equipment from Columbus Watch Co.

196. Cornell Watch Co. #15,381 ca. 1871.

Dueber Hampden — see Hampden.

Edgemere Watch Co. — name used by Sears Roebuck & Co. on watches made by Seth Thomas.

Elgin — see National Watch Co.

Empire City Watch Co. — see United States Watch Co. of Marion, NJ.

Equity Watch Co. — name used on watches by Waltham Watch Co.

Fredonia Watch Co. — see Independent Watch Co.

Freeport Watch Co. — 1874-1875, Freeport, IL

This was an enterprise built around watches made by Mr. C. H. Hoyt for M. S. Smith jewelers in Detroit. These were high quality, of Hoyt's design, and built with interchangeable parts on Hoyt's own equipment. Hoyt's operation was purchased by Eber Ward of Detroit and established in Freeport along with machinery from Don Mozart's defunct company in Ann Arbor. A fire destroyed the building before any progress was made but a few pieces marked Freeport were finished from Hoyt's material.

Gruen Watch Co.[5]

1874-1879	D. Gruen	Columbus, OH
1879-1882	Gruen & Savage	Columbus
1894-1898	D. Gruen & Son	Columbus
1898-1922	D. Gruen & Sons	Cincinnati
1922-1955	Gruen Watch Co.	Cincinnati
1955-	Gruen Industries Inc.	New York, NY

Dietrich Gruen began with a business finishing special order Swiss ebauches, expanding to the partnership of Gruen & Savage. The latter formed the nucleus of the Columbus Watch Company, which Gruen left in 1894 to resume dealing in imported movements. A new partnership with his son began with special order movements made by Assmann in Dresden, Germany. Though of exceptional quality, these were dropped around 1904 for Swiss movements available in larger quantity and thinner styles. As business expanded, Gruens acquired manufacturing facilities in Switzerland, supplying ebauches to ever growing finishing

197. Factory of Freeport Watch Co., Freeport, Illinois — burned in 1875.

shops in Cincinnati. The operation remained a family-owned business until 1955.

Hamilton Watch Co. — 1892-1970's, Lancaster, PA

Hamilton was assembled with machinery from the defunct Keystone Standard Watch Co. and Aurora Watch Company. It was to have marketed watches in 1892 (400th anniversary of Columbus' discovery of America) as the Columbian Watch Co. and their first movement was to have been the "Hamilton." Since the name Columbia Watch Co. was already protected, the company name became Hamilton Watch Company.[6] They apparently considered manufacturing in Aurora, IL, but consolidated in Lancaster, PA, in 1892. Their production before 1900 included some 7- to 15-jewel watches, but thereafter was limited to movements of 17 jewels or more, and better than average quality. With this, they maintained a sound reputation, remaining popular in the rail trade.

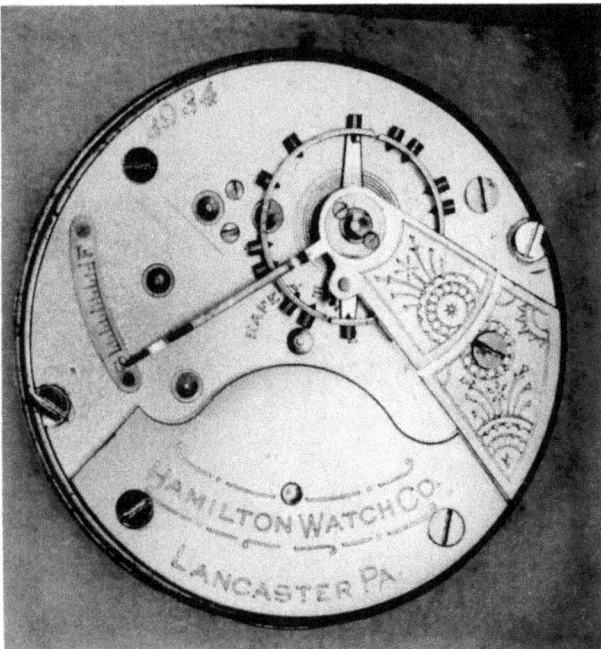

198. Hamilton Watch Co. 7-jewel movement #2934 ca. 1894.

199. Hampden railroad movement #3,268,395 ca. 1910.

Hampden Watch Co.[7]

	Don Mozart formed	
1864-1875	New York Watch Co.	Providence, RI
	1867 moved to	Springfield, MA
1875-1876	New York Mfg. Co.	Springfield
1877-1888	Hampden Watch Co.	Springfield
	purchased by Dueber Watchcase Co.	
1889-1923	Hampden Watch Co.	Canton, OH
1923-1930	Dueber Hampden Watch Co.	
		Canton

After Don Mozart left, the New York Watch Co. became an early producer of high quality ¾-plate watches. Because of financial difficulties this reorganized, eventually becoming Hampden, named after Hampden County where Springfield is located. Merger with Dueber Watchcase Co. resulted in relocation to Ohio and this was eventually bought by Russia to establish a complete watch business there.

Home Watch Co. — name used on inexpensive watches by American Watch Co.

Howard Watch Co.

	separated from Boston Watch Co. in 1857	
1857-1858	Howard & Rice	Roxbury, MA
1858-1861	E. Howard & Co.	Roxbury
1861-1863	Howard Clock & Watch Co.	Roxbury
1863-1881	Howard Watch & Clock Co.	Roxbury
1881-1903	E. Howard Watch & Clock Co.	Roxbury
1903-ca. 1930	Keystone Watchcase Co.	Waltham, MA

After failure of the Boston Watch Co. Edward Howard re-established himself at his factory in Roxbury. He kept production low and concentrated on high quality. While experimenting considerably, Howard's movement styles were conservative, keeping with low jewel counts and flat hairsprings through most of the 19th Century. After being purchased by Keystone, Howard watches made at Waltham remained high in quality and movement designs were updated to contemporary styles. Keystone had bought only the watchmaking part of the Howard company, establishing itself in the newly purchased factory that had previously been the United States Watch Co. of Waltham. By 1927 business was slow and soon thereafter the E. Howard Clock Co., then still in Roxbury, purchased the Waltham factory from Keystone, where it remains in business today.

Howard Davis & Dennison — see American Watch Co.

Illinois Watch Co.

1869-1879	Illinois Springfield Watch Co.	Springfield, IL
1879-1885	Springfield Illinois Watch Co.	Springfield
1885-1927	Illinois Watch Co. sold to Hamilton	Springfield

This was the only new company to enter watch manufacturing between 1864 and 1874. Illinois appealed to the railroad market and maintained a sound business. Though production never approached Waltham or Elgin, they were the third largest producer of conventional-jeweled watches behind those two companies.

Independent Watch Co.[13]

1880-1883	Independent Watch Co.	Fredonia, NY
1883-1885	Fredonia Watch Co.	Fredonia
1885-1895	Peoria Watch Co.	Peoria, IL

This was the outgrowth of a catalogue business run by the Howard brothers of Fredonia (no known relation to Edward Howard). They concentrated on patent medicines

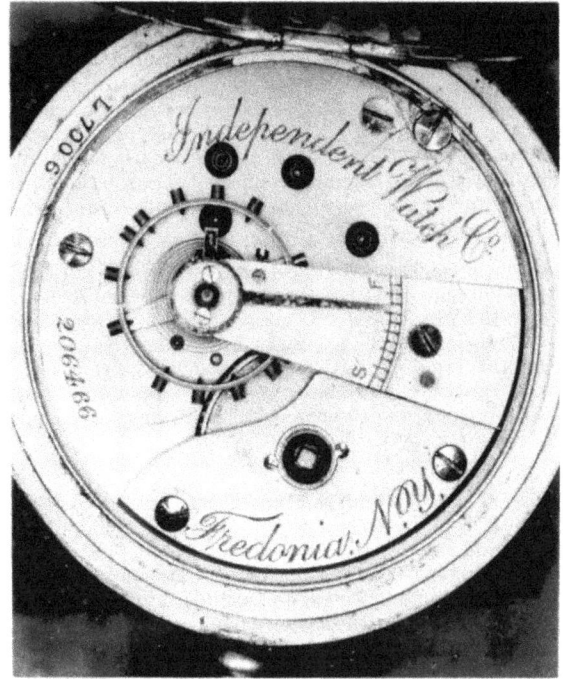

201. Independent Watch Co., Fredonia, New York, #206,-466 ca. 1883. This used U.S. Marion plates and train, and the serial number is probably a U.S. Marion number. Later Fredonia Watch Co. movements combined the U.S. Marion straight line lever train with Newark plates (illustration 166).

200. Illinois Watch Co. factory at Springfield, Illinois, ca. 1890.

but also sold American and Swiss watches. Watches became a large enough business that they decided to make their own, purchasing equipment from the defunct Newark and U.S. Marion companies. There were always technical difficulties plus merchandising problems because of resistance from the jewelry trade. After a brief struggle they sold their machinery, forming the Peoria Watch Company, to which they were financially connected. Peoria remained in business for some years, appealing to the railroad and moderate-priced market.

Ingersoll, Robert H. & Bro.[8] — 1881-1922, New York, NY

Ingersoll merchandised the inexpensive unjeweled watch, driving the price to $1. Their first watches were bought from Waterbury Clock Co. but after mammoth success they bought the Trenton and New England Watch companies to increase production. Business became shaky in the early 1920's, probably due to overcompetition, and the Ingersolls sold to Waterbury Clock Co.

Ingraham

| 1905-1911 | Bannatyne Watch Co. | Waterbury, CT |
| 1911-1971 | E. Ingraham | Bristol, CT |

Archibald Bannatyne developed his own dollar watch and company in 1905. The E. Ingraham Company, founded in 1835, bought this in 1911 and expanded into one of the major producers. When they were bought by McGraw Edison in 1967 they were still making a watch virtually identical to Bannatyne's original model.

International Watch Co. — 1902-1907, Newark, NJ

This was a minor producer of dollar watches during the early part of this century. They are not to be confused with the International Watch Co. of Schaffhausen, Switzerland.

Keystone Standard Watch Co. — see Lancaster Watch Co.

Keystone Watch Case Co. — 1885-1927, Philadelphia, PA

Through expansion and purchase, Keystone gained control of:

> Howard Watch Co.
> New York Standard Watch Co.
> Crescent Watch Case Co.
> Philadelphia Watch Case Co.

and through purchase, terminated the lives of:

> United States Watch Co. of Waltham
> Suffolk Watch Co. of Waltham.

Knickerbocker Watch Co. — 1890-1930, New York, NY

Knickerbocker has a somewhat obscure history, selling Swiss imports and American duplex watches. In this last regard they appear to have been dealing in watches by New England Watch Co. marked with the Knickerbocker name. In the early 1920's the company was bought by young Ira Guilden, who later became vice-president of Bulova and president of Waltham.

Lancaster Watch Co.

1874-1876	Adams & Perry	Lancaster, PA
1877-1878	Lancaster Watch Co.	Lancaster
1878-1879	Lancaster Penn. Watch Co.	Lancaster
1879-1886	Lancaster Watch Co.	Lancaster
1886-1890	Keystone Std. Watch Co.	Lancaster

Lancaster is the best known name in a series of companies existing in Lancaster before Hamilton. Adams & Perry intended to make high-grade ¾-plate watches but never got into production. The succeeding companies made average-grade movements but could not survive in the highly competitive watch industry. The Keystone Standard equipment was finally bought by the founders of Hamilton.

Manhattan Watch Co. — 1883-1892, New York, NY

This short-lived company was a minor producer of inexpensive jeweled watches. These were unique movements

202. Lancaster Watch Co. factory in Lancaster, PA.

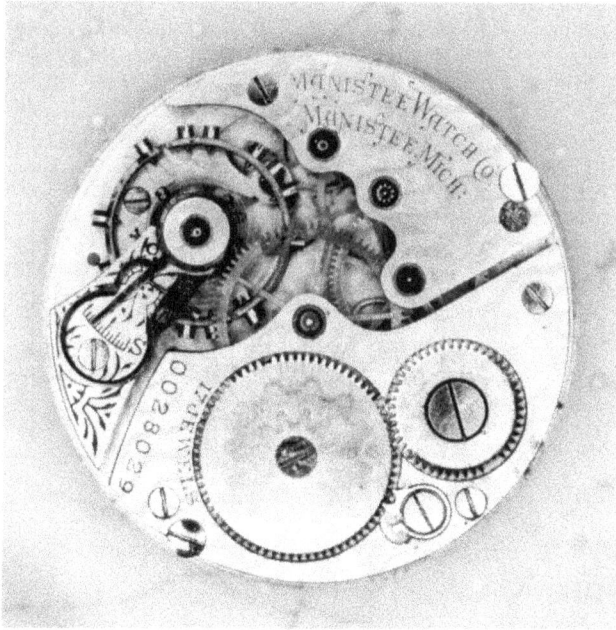

203. Manistee Watch Co. movement #28,029 ca. 1910.

with unusual escapements and setting mechanisms. Some center-sweep models were also made.

Manistee Watch Co.[9] — 1908-1912, Manistee, MI

To Manistee goes the honor of being the last watch company formed in the United States. They made inexpensive jeweled watches, quickly failing under pressure from the large factories since this was the most competitive area of the industry.

Marion Watch Co. — see United States Watch Co. of Marion

McIntyre Watch Co.[10] — ca. 1915, Kankakee, IL

This was an obscure company which never reached production. Charles DeLong made several prototype watches for them, and there was perhaps a small amount of watch material and tooling begun.

Melrose Watch Co. — see Tremont Watch Co.

Mozart Watch Co.[7, 26]

	Mozart formed	
1864	New York Watch Co.	Providence, RI
	he was discharged and then formed	
1866-1870	Mozart Watch Co.	Ann Arbor, MI
	equipment sold to	
1871-1872	Rock Island Watch Co.	Rock Island, IL
	equipment sold to	
1874-1875	Freeport Watch Co.	Freeport, IL

After leaving the New York Watch Co. Don Mozart formed a new venture in Ann Arbor to make his unique watch. This failed after a few (perhaps 50) pieces were finished. With Mozart's machinery the Rock Island Watch Co. intended to make Mozart's watch but never got started, and the machinery finally went to the Freeport Watch Company.

Nashua Watch Co. — 1859-1862, Nashua, NH

Nashua was one of the earliest companies. They were aggressive in attempting complete interchangeability, most advanced in machine design, and earliest to introduce ¾-

plate movements. Before reaching production Nashua was purchased by the American Watch Company, a great technical advance for Waltham.[11, 12]

National Watch Co.

| 1864-1874 | National Watch Co. | Elgin, IL |
| 1874-1960's | Elgin National Watch Co. | Elgin |

Known simply as Elgin, this was the other large company beside Waltham, eventually surpassing Waltham in production. Compared to Waltham they manufactured fewer basic movement styles, with less ostentatious finish. They made watches of all grades and their finest were very high quality.

Newark Watch Co.[13] — 1864-1870, Newark, NJ

Newark was one of the wave of companies in 1864. The company and its watch were never successful. After going out of business their machinery was bought by Cornell Watch Co. which ushered the old Newark watch along a series of failures. See Cornell Watch Co.

New England Watch Co. — see Waterbury Watch Co.

New Haven Clock Co. — 1853-1946, New Haven, CT, began watches in 1880.

This was one of the big three dollar watch makers behind Ingersoll and Westclox, and also one of the longest lived. They started with a large back-wind watch, a sort of alarm clock derivative, then refined this to a typical dollar watch type of movement with pin pallet escapement.

New Haven Watch Co. — see Trenton Watch Co.

New York Chronograph Watch Co. — name used by Manhattan Watch Co.

New York City Watch Co. — 1890-1897, New York, NY

This firm was a small dollar watch manufacturer that made the interesting crank wind watch. What appeared to be a winding stem was folded over sideways to wind the mainspring.

New York Mfg. Co. — see Hampden.

New York Standard Watch Co. — 1885-1929, Jersey City, NJ

The unusual worm-drive escapement was among the gimmicks used by New York Standard for attracting attention to their first watch. They reverted to a more typical design to become a notable maker of inexpensive jeweled watches,

204. Early view of National Watch Co. in rural Elgin, Illinois, ca. 1870.

205. Otay Watch Co. #1411 ca. 1890.

---

as indicated by their long lifespan. They were acquired by Keystone Watch Case Co. in 1903.

New York Watch Co. — see Hampden Watch Co.

New York Watch Co. — a name used by the New York City Watch Co. — ca. 1895.

Non-Magnetic Watch Co. of America[14] — ca. 1887-ca. 1900.
This company marketed watches with Paillard's patent non-magnetic balances and hairsprings. Fullplate 18-size movements were made by Peoria and 16-size bridge movements were Swiss, reportedly made by Aeby & Co. and J. J. Badollet. Little is known of the sales organization or how it operated.

Non-Magnetic Watch Co. — *Chicago* — name used on watches made by Illinois using Paillard's patent balances and hairsprings. There was a full line of non-magnetic watches in fullplate and bridge models.

Otay Watch Co.[15]

1889-1890	Otay Watch Co.	Otay, CA
1891	San Jose Watch Co.	San Jose, CA

Otay was formed near San Diego under the urging of a family named Kimbal, with P. H. Wheeler the watchmaking brain. Wheeler had previously been with U.S. Marion, Rockford, Illinois, and Columbus, holding numerous patents. A small output of decent watches was made before failing, and the equipment was sailed up to San Jose to start another watch venture. Few known watches were produced at San Jose and the machinery went to Osaka, Japan, where a few examples were manufactured, identical to the Otay watch.

Peoria Watch Co.[13] — 1885-1895, Peoria, IL
This watch company was assembled with machinery from the Fredonia company. Both the machinery and watch were mergers of Newark and U.S. Marion products. Peoria dealt mostly with the railroad trade and also made 18-size fullplate watches marked Non-Magnetic Watch Co. of America.

Philadelphia Watch Co.[16] — 1868-late 1870's, Philadelphia, PA
This mysterious company was something of a contractor and sales organization similar to Webb C. Ball. It was run by Eugene Paulus, who had movements of his plate design made in Switzerland. These ranged in sizes and quality, including many well made watches.

Pitkin Brothers[17] — 1838-1842, Hartford, CT
The Pitkin brothers were the first Americans to attempt interchangeable finished watch parts. They found moderate technical success but failed as a business. Were it not for their business difficulties they might have spawned the American watch industry before Aaron Dennison.

Plymouth Watch Co. — name used by Sears Roebuck & Co. on watches made by Rockford Watch Company.

Remington Watch Co. — see Appleton Watch Co.

Rockford Watch Co.

1874-1896	Rockford Watch Co.	Rockford, IL
1896-1915	Rockford Watch Co. Ltd.	Rockford

One of the few companies to last 40 years was Rockford. They maintained a business in the railroad trade but also offered average and lower grade models. Production was never high and on the average they manufactured in two weeks what Elgin made in a day.

Rock Island Watch Co. — see Mozart Watch Co.

Royal Gold American Watch, New York — name used by U. S. Marion, ca. 1876.

San Jose Watch Co. — see Otay Watch Co.

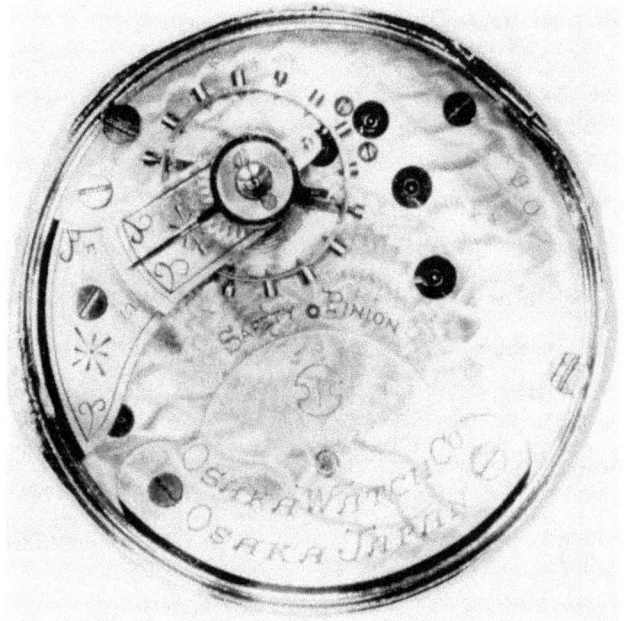

206. Osaka Watch Co. #904 ca. 1895 — made with old Otay Watch Co. machinery.

132

207. Peoria Watch Co. factory in Peoria, Illinois, ca. 1890.

Santa Fe Watch Co. — name used by a sales company on watches made by Illinois.

Seth Thomas Clock Co. — 1813-present, Plymouth, CT, watches 1884-1914.

Seth Thomas entered watch production as a large company able to maintain significant manufacturing rates, thus becoming America's fifth largest pocket watch maker. Their first model was a ¾-plate with long lever and they eventually offered a full range of movement grades and styles. They were prone to dazzling customers with high jewel counts in their high-grade models, but their primary business was inexpensive jeweled watches in the original ¾-plate layout. Hard competition drove them from business during the period when Ingersoll brothers dominated low-cost pocket watches and wristwatches were rising in popularity.

South Bend Watch Co.[18] — see Columbus Watch Co.

Springfield Illinois Watch Co. — see Illinois Watch Co.

Sterling Watch Co. — name used by E. Ingraham.

Stevens, J. P. — see Bowman.[19]

Suffolk Watch Co. — see Columbia Watch Co.[20]

Tracy Baker & Co. — see American Watch Co.

Tremont Watch Co.

| 1864-1866 | Tremont Watch Co. | Boston, MA |
| 1866-1868 | Melrose Watch Co. | Melrose, MA |

Aaron Dennison put together the Tremont Co. as an attempt to avoid expensive American labor. Escapements, balances, geartrains and fine parts were made in Switzerland then mated with heavy brass and steel parts made in Boston. This was a successful business, and the owners concluded to build the entire watch, moving to the Boston suburb of Melrose. The Melrose Watch Co. got out some watches but expenses rapidly overran financial resources and the operation closed.

Trenton Watch Co.

1883-1887	New Haven Watch Co.	New Haven, CT
1887-1907	Trenton Watch Co.	Trenton, NJ
1908-1922	Ingersoll Trenton	Trenton

This operation started out as the New Haven Watch Co. but after a short period relocated and renamed the plant. They made inexpensive jeweled watches, continuing in this line after being purchased by Ingersoll brothers.

Union Watch Co.[21] — 1875-1878, Fitchburg, MA

After failure of the Marion Watch Co. Sylvanus Sawyer, a stockholder, moved some of the equipment to his hometown of Fitchburg. Little came of this effort to make watches and the operation closed without getting into production.

United States Watch Co.[22] of Marion, NJ, (Jersey City)

1864-1872	United States Watch Co.	Marion, NJ
1872-1874	Marion Watch Co.	Marion
1874-1877	Empire City Watch Co.	Marion

One of the companies of 1864 to manufacture a truly good watch was United States. They were first in America with stemwinding, damaskeening, and were one of the early companies to introduce a high-grade ¾-plate movement. Overexpenditure on the business and facilities placed them in an unsound financial position and they closed.

United States Watch Co. of Waltham[23] — 1884-1896, Waltham, MA

After leaving American, Charles Vander Woerd started the United States Watch Co. of Waltham. They made several interesting models of moderate quality but could never meet the competition of the large producers. They absorbed the Suffolk Watch Co. of Waltham and later became one of the companies bought by Keystone Watch Case Company. Their factory building was then used to make the Keystone Howard watches till approximately 1930 when it was bought by E. Howard Clock Company, who is still there today.

Waltham Watch Co. — see American Watch Co.

Warren Manufacturing Co. — see American Watch Co.

Washington Watch Co.[24] — 1874-1875, Washington, DC

This company was started by Jason Hopkins to manufacture a duplex escapement watch but never reached production. Very little is known of the company or its product.

Washington Watch Co.[24]

For a time, Montgomery Ward used this name on watches made by the Illinois Watch Company.

208. Seth Thomas Henry Molineux model #55,120 ca. 1888.

Waterbury Clock Co. — 1857-1944, Waterbury, CT, began
 watches 1890.

Waterbury Clock started in the clock business as a
branch of Benedict & Burnham Mfg. Company. They introduced a clock-like watch in 1890, then in 1894 began supplying cheap watches to the Ingersoll brothers. Waterbury
Clock bought the Ingersoll business and Ingersoll name in
1922, then became part of the organization that created
U.S. Time Corp. in 1944, makers of Timex.

Waterbury Watch Co.

1833-1880	Benedict & Burnham	Waterbury, CT
	began making watches in 1878	
1880-1898	Waterbury Watch Co.	Waterbury
1898-1914	New England Watch Co.	Waterbury

Waterbury launched into business on the longwind duplex watch of D. A. Buck, and launched the American dollar watch industry. Over their career they dropped the
longwind design and added inexpensive jeweled watches to
their catalogue, still with the duplex escapement. The business continued with Waterbury designs after reorganizing
as the New England Watch Company. Ingersoll brothers
acquired the company for dollar watch production, but even
with their lifespan thus foreshortened they were the fourth
largest dollar watch maker as of 1930.

Western Clock Co. — Westclox — 1887-present, Peru, IL,
 first watch 1899.

The Western Clock Co. began in 1887 by taking over the
defunct United Clock Company. The latter was begun in
1885 by several men from Waterbury, Connecticut. Western
Clock Co. became Westclox, began watch production around
the turn of the century, and was a major dollar watch producer. Today they are America's only pocket watchmaker,
representative of America's two greatest contributions to
watch technology: the inexpensive watch, and machine production of interchangeable parts.

Western Watch Co. — see Cornell Watch Co.

Wichita Watch Co. — 1887, Wichita, KS

No examples are known of this company's product. They
apparently completed a factory building and had plans for
an 18-size half-plate watch. The scheme must have run

short of money soon thereafter for little else is known of
their activity.

Williamstown Watch Co. — 1883-1885, Williamstown, MA

This was the first venture of Robert J. Clay to promote
his worm drive escapement, later used by New York Standard. Financial backers declined to become involved with
such a design and a more conventional watch was proposed.
An old twine factory was outfitted for watchmaking and
twelve thousand movements begun. Squabbles and lawsuits
among stockholders and managers brought work to a stop
so that the enterprise eventually dissolved.

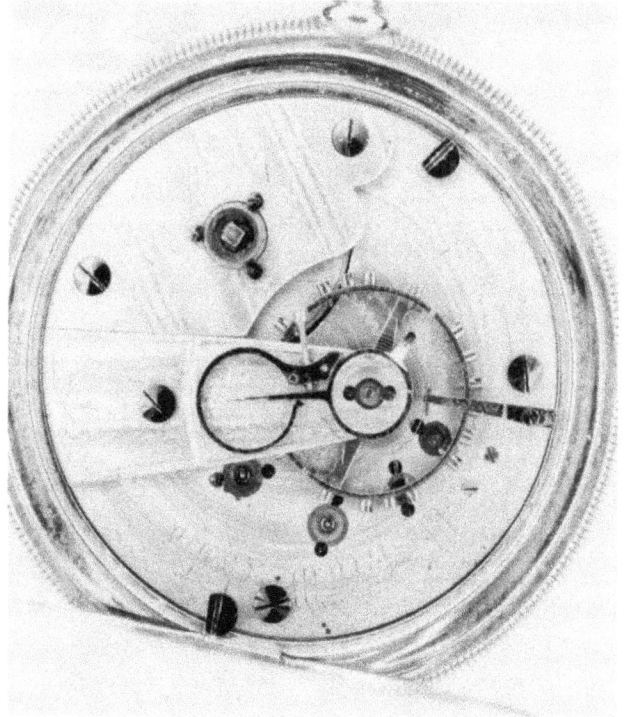

209. Western Watch Co. #31,999 ca. 1880, a direct descen-
dant of the Newark watch.

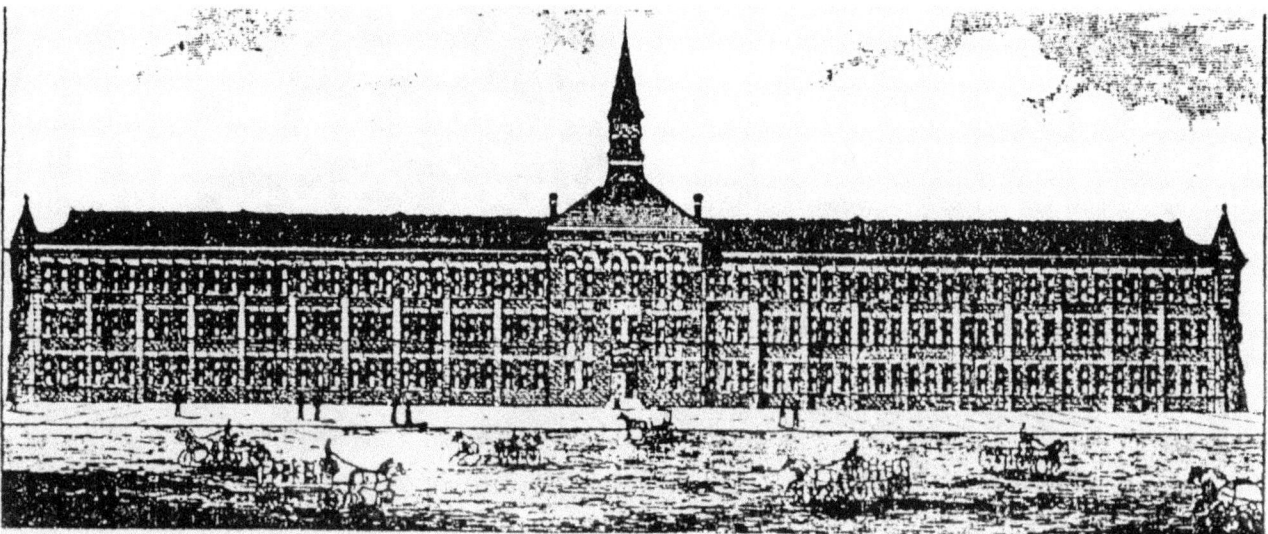

210. Wichita Watch Co. building ca. 1887 — possibly an artist's depiction since the company never reached production.

REFERENCES

General information on American watch companies may be found in the following:

a. Henry G. Abbott, *Watch Factories of America* (reprinted Adams Brown Co., Exeter, NH).

b. Charles S. Crossman, *The Complete History of Watchmaking in America* (reprinted Adams Brown Co., Exeter, NH).

c. George E. Townsend, *Almost Everything You Wanted to Know About American Watches and Didn't Know Who To Ask* (published Alma, MI, George Townsend).

d. George E. Townsend, *American Railroad Watches* (published Alma, MI, George Townsend).

e. George E. Townsend, *The Watch That Made The Dollar Famous* (published Alma, MI, George Townsend).

Large numbers of photographs may be found in annual *American Pocket Watch Price Indicators* published by Roy Ehrhardt.

1. Charles W. Moore, *Timing A Century* (Harvard University Press, Cambridge, MA, 1945).

2. Henry G. Abbott, *A Pioneer* (reprinted Adams Brown Co., Exeter, NH, 1968).

3. James W. Peghiny, *Some Interesting Photographs of the Auburndale Watch Company* (BULLETIN, National Association of Watch and Clock Collectors, Inc., December 1970), p. 790.

4. Stoltz & Parkhurst, *William Wallace Dudley and His "Masonic Watch"* (BULLETIN, National Association of Watch and Clock Collectors, Inc., October 1968), p. 496.

5. Eugene T. Fuller, *The Priceless Possession of a Few* (BULLETIN, National Association of Watch and Clock Collectors, Inc., Supplement, Winter 1974).

6. Donald Hoke, *The First Aurora Watch* (BULLETIN, National Association of Watch and Clock Collectors, Inc., August 1977), p. 331.

7. W. Barclay Stephens, *The New York Watch Co.* (BULLETIN, National Association of Watch and Clock Collectors, February 1951), p. 279.

8. Robert F. Tschudy, *Ingersoll "The Watch That Made The Dollar Famous"* (BULLETIN, National Association of Watch and Clock Collectors, April 1952), p. 97.

9. George E. Townsend, *The Manistee Watch Co.* (BULLETIN, National Association of Watch and Clock Collectors, Inc., February 1976), p. 29.

10. Paul M. Chamberlain, *It's About Time* (New York, Richard R. Smith, 1941), p. 471.

11. Frederick M. Selchow, *Belding Dart Bingham — The Nashua Watch Co.* (BULLETIN, National Association of Watch and Clock Collectors, Inc., December 1975), p. 555.

12. Thomas De Fazio, *The Nashua Venture and The American Watch Company* (BULLETIN, National Association of Watch and Clock Collectors, Inc., December 1975), p. 581.

13. W. Barclay Stephens, *The Newark Watch Co. and Its Career* (BULLETIN, National Association of Watch and Clock Collectors, February 1950), p. 73.

14. Thomas De Fazio, *A Note Concerning the Non-Magnetic Watch Company of America* (BULLETIN, National Association of Watch and Clock Collectors, Inc., February 1975), p. 39.

15. Chamberlain, p. 279.

16. Larry Treiman, *The Philadelphia Watch Company — Revisited* (BULLETIN, National Association of Watch and Clock Collectors, Inc., December 1978), p. 597.

17. Dr. P. L. Small, *The Pitkin Brothers* (BULLETIN, National Association of Watch and Clock Collectors, October 1954), p. 251.

18. Paul Berg, *The South Bend Watch Co.* (BULLETIN, National Association of Watch and Clock Collectors, Inc., August 1971), p. 1184.

19. W. B. Stephens, *J. P. Stevens — His Watch Company and Inventions* (BULLETIN, National Association of Watch and Clock Collectors, December 1954), p. 321.

20. Edmund L. Sanderson, *Waltham Industries* (Waltham, MA, Waltham Historical Society, 1957), pp. 106, 108.

21. Frederick M. Selchow, *The Watch Company of Fitchburg, Massachusetts* (BULLETIN, National Association of Watch and Clock Collectors, Inc., April 1969), p. 914.

22. W. Barclay Stephens, *The United States Watch Company of Marion, New Jersey* (BULLETIN, National Association of Watch and Clock Collectors, June 1950), p. 148.

23. Sanderson, p. 105.

24. A. E. Mathews, *The Washington Watch Company* (BULLETIN, National Association of Watch and Clock Collectors, Inc., April 1973), p. 1094.

25. Lawrence W. Treiman, *Railroad Watches and Time Service* (BULLETIN, National Association of Watch and Clock Collectors, Inc., October 1972), p. 651.

26. David C. Olson, *The Mozart Watch* (BULLETIN, National Association of Watch and Clock Collectors, Inc., June 1957), p. 548.

APPENDIX B
SERIAL NUMBERS AND DATES

The following tables show approximate production quantities for given dates through the industry. This data has been fairly well determined for major manufacturers, but for lesser companies and dollar watch producers the figures are approximate. In these instances, quantities were estimated from reported production rates, known watches and occasional bits of information. It should be remembered that lists such as these cannot be exact.

COMPANY	1860	1865	1870	1875	1880	1885	1890	1895	1900	1905	1910	1915	1920	1925	1930
Waltham	20K	190K	450K	900K	1.5M	2.6M	4.4M	6M	9M	13M	17M	20M	23M	25M	27M
Howard*	3K	50K	73K	200K	300K	430K	600K	700K	800K	900K	1M	1.3M	1.4M	1.45M	1.5M
U.S. Marion			40K	250K											
Newark			7K	13K											
Elgin			100K	300K	700K	1.8M	4M	6M	9M	12M	15M	18.5M	23M	28M	33M
Hampden				40K	200K	450K	700K	1.0M	1.3M	2.8M	3.3M	3.8M	4.3M	4.6M	
Illinois				80K	190K	280K	430K	600K	1M	2.5M	3.8M	4.2M	4.4M	5.2M	
Lancaster					50K	100K	150K								
Rockford				6K	50K	100K	165K	260K	410K	580K	760K	935K			
Auburndale					3K	6K									
Waterbury					100K	1M	3M	5M	7.5M	9M	12M				
Peoria								50K	75K	100K					
New Haven Clock							2M	3M	5M	7M	10M	14M	18M	25M	30M
Columbus / South Bend						120K	175K	350K	425K	500K	600K	750K	930K	1.1M	1.2M
Cheshire							50K	100K							
Trenton*							150K	500K	2M	2.5M	3M	3.5M	4M		
Manhattan						200K	500K								
Aurora						150K									
Seth Thomas						4K	110K	290K	500K	760K	1.3M	3.6M			
U.S. Waltham							10K	150K	225K	450K					
NY Standard							600K	900K	1.2M	1.5M	1.8M	2.1M	2.4M	2.7M	3M
Hamilton								5K	100K	400K	1.2M	1.5M	1.8M	2.2M	2.5M
Ingersoll								1.2M	6M	10M	25M	40M	54M	66M	74M
Westclox										4M	10M	18M	25M	37M	50M
Ansonia											1M	3M	5M	7M	10M
E. Ingraham										50K	250K	2M	4M	7M	10M
Manistee											40K	60K			
COMPANY	1860	1865	1870	1875	1880	1885	1890	1895	1900	1905	1910	1915	1920	1925	1930

K = Thousand
M = Million

YEAR

*Serial numbers were not manufactured sequentially, making them difficult to date.

SUBJECT INDEX

137

BIOGRAPHICAL INDEX

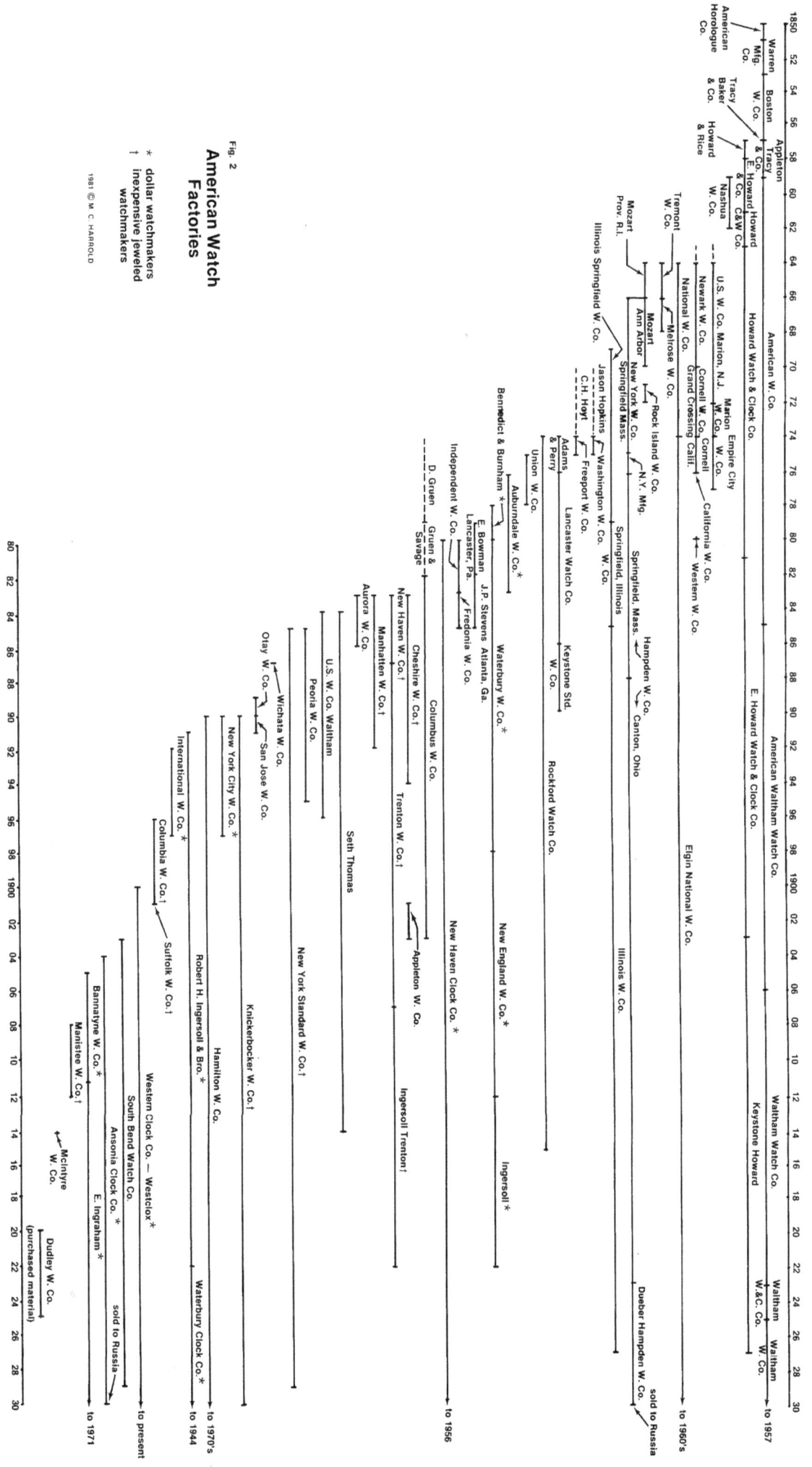

American Watch Factories

Fig. 2

1850 52 54 56 58 60 62 64 66 68 70 72 74 76 78 80 82 84 86 88 90 92 94 96 98 1900 02 04 06 08 10 12 14 16 18 20 22 24 26 28 30

1981 © M C HARROLD

* dollar watchmakers
† inexpensive jeweled watchmakers

American Horologue Co.
Warren Mfg. Co.
Boston W. Co.
Tracy Baker & Co.
Howard & Rice
Appleton Tracy & Co.
E. Howard & Co.
Howard & Co.
E. Howard & Co.
C&W Co.
Nashua W. Co.
Tremont W. Co.
Mozart Prov. R.I.
Mozart Ann Arbor
Melrose W. Co.
National W. Co.
Newark W. Co.
U.S. W. Co. Marion, N.J.
Marion W. Co.
Cornell Co. Cornell
Grand Crossing Calif.
Cornell
California W. Co.
Western W. Co.
New York W. Co.
Rock Island W. Co.
N.Y. Mfg.
Empire City
Howard Watch & Clock Co.
American Watch Co.
American Waltham Watch Co.
Illinois Springfield W. Co.
Springfield Mass.
Springfield, Illinois
Springfield, Mass.
Jason Hopkins
Washington W. Co.
C.H. Hoyt
Freeport W. Co.
Adams & Perry
Union W. Co.
Lancaster Watch Co.
Keystone Std. W. Co.
Hampden W. Co.
Canton, Ohio
Illinois W. Co.
Benedict & Burnham *
Auburndale W. Co. *
E. Bowman Lancaster, Pa.
J.P. Stevens Atlanta, Ga.
Fredonia W. Co.
Waterbury W. Co. *
Rockford Watch Co.
Independent W. Co.
D. Gruen
Gruen & Savage
Aurora W. Co.
Manhatten W. Co. †
New Haven W. Co. †
Cheshire W. Co. †
Columbus W. Co.
Trenton W. Co. †
Appleton W. Co.
New Haven Clock Co. *
New England W. Co. *
Ingersoll *
Otay W. Co.
Wichata W. Co.
San Jose W. Co.
Peoria W. Co.
U.S. W. Co. Waltham
New York City W. Co. *
International W. Co. *
Columbia W. Co. †
Suffolk W. Co. †
Seth Thomas
New York Standard W. Co. †
Robert H. Ingersoll & Bro. *
Knickerbocker W. Co. †
Bannatyne W. Co. *
Manistee W. Co. †
McIntyre W. Co.
Western Clock Co. — Westclox *
South Bend Watch Co.
Ansonia Clock Co. *
E. Ingraham *
Hamilton W. Co.
Ingersoll Trenton †
Appleton W. Co.
New England W. Co. *
Ingersoll *
Dudley W. Co.
(purchased material)
Dueber Hampden W. Co.
Elgin National W. Co.
E. Howard Watch & Clock Co.
Keystone Howard
Waltham Watch Co.
Keystone Howard
Waltham W.&C. Co.
Waltham W. Co.
Waterbury Clock Co. *

sold to Russia
sold to Russia
to 1971
to present
to 1944
to 1970's
to 1956
to 1957
to 1960's

www.ingramcontent.com/pod-product-compliance
Lightning Source LLC
Chambersburg PA
CBHW080556220326
41599CB00032B/6499